SpringerBriefs in Electrical and Computer Engineering

For further volumes:
http://www.springer.com/series/10059

Chulwoo Kim • Hyun-Woo Lee
Junyoung Song

High-Bandwidth Memory Interface

Chulwoo Kim
Department of Electrical Engineering
Korea University
Seongbuk-gu, Seoul
Korea

Junyoung Song
Department of Electrical Engineering
Korea University
Seongbuk-gu, Seoul
Korea

Hyun-Woo Lee
SK-Hynix
Gyeonggi-do
Korea

ISSN 2191-8112 ISSN 2191-8120 (electronic)
ISBN 978-3-319-02380-9 ISBN 978-3-319-02381-6 (eBook)
DOI 10.1007/978-3-319-02381-6
Springer New York Heidelberg Dordrecht London

Library of Congress Control Number: 2013950761

© The Author(s) 2014

This work is subject to copyright. All rights are reserved by the Publisher, whether the whole or part of the material is concerned, specifically the rights of translation, reprinting, reuse of illustrations, recitation, broadcasting, reproduction on microfilms or in any other physical way, and transmission or information storage and retrieval, electronic adaptation, computer software, or by similar or dissimilar methodology now known or hereafter developed. Exempted from this legal reservation are brief excerpts in connection with reviews or scholarly analysis or material supplied specifically for the purpose of being entered and executed on a computer system, for exclusive use by the purchaser of the work. Duplication of this publication or parts thereof is permitted only under the provisions of the Copyright Law of the Publisher's location, in its current version, and permission for use must always be obtained from Springer. Permissions for use may be obtained through RightsLink at the Copyright Clearance Center. Violations are liable to prosecution under the respective Copyright Law.

The use of general descriptive names, registered names, trademarks, service marks, etc. in this publication does not imply, even in the absence of a specific statement, that such names are exempt from the relevant protective laws and regulations and therefore free for general use.

While the advice and information in this book are believed to be true and accurate at the date of publication, neither the authors nor the editors nor the publisher can accept any legal responsibility for any errors or omissions that may be made. The publisher makes no warranty, express or implied, with respect to the material contained herein.

Printed on acid-free paper

Springer is part of Springer Science+Business Media (www.springer.com)

Preface

These days most people use at least one personal computing device, like a pc, a tablet or a smartphone. DRAM is one of the data storage components in these devices, and memory bandwidth has increased significantly over the last 10 years. Main memories like DDR2, DDR3 and DDR4 are used in PCs and servers. DDR4 is a higher-bandwidth successor to DDR2 and DDR3 to support high-speed applications. Graphic DDR (GDDR) is used in graphic cards and game consoles. GDDR series provide higher-bandwidth than main memories to process large image and video data such as frame data buffering and life-like 3D image rendering.

Many high-speed wireline interface techniques for other applications have been adopted in SDRAM to increase the bandwidth. However, there are several key differences between memory interface and other wireline applications, which cause difficulties for synchronous DRAM (SDRAM) design. First, SDRAM cannot utilize differential signaling due to inherently large number of data pins. Second, a DDR memory module consists of multi rank and multi DIMM which causes impedance discontinuities due to many stubs, connectors, vias and so on. Therefore, the data channel of DRAM is more vulnerable to noise coupling and inter symbol interference. Third, simultaneous switching output noise is a problem because DQ drivers or receivers usually operate all at the same time. The last major difference of high-bandwidth DRAM design is that the transistor performance of DRAM process is very poor compared to that of logic process.

There are many technologies to enhance the memory interface such as TSV interface, high-speed serial interface including equalization, ODT, pre-emphasis, wide I/O interface including crosstalk, skew cancellation, and clock generation and distribution. This book provides overviews of recent advances in memory interface design both in architecture and at circuit levels. Subtopics will include signal integrity and testing. Future trends for further bandwidth enhancement will be covered as well. This book is developed mainly from the research outputs on memory interface by the Advanced Integrated System Lab of Korea University and a number of notable research papers from other groups. This book provides readers with crucial resources to develop their studies about high bandwidth memory systems.

Chapter 1 investigates the basic DRAM characteristics including cell array access control, read and write operation, and an example of DDR3 configuration. Power dissipation is a big concern in this mobile era. The important considerations for active and stand-by power consumptions are described. Chapter 2 deals with I/O configuration for DDR SDRAM. It provides the readers with an understanding for speed limitation of cell access. Several important DRAM organizations are compared to differentiate and to help understand the various types of DRAM. Chapter 3 focuses on the design of clock generation and distribution. Because DLL (Delay Locked Loop) is widely used in DDR SDRAM, in-depth investigation and analysis of DLL is provided. To improve the data valid window for high-speed DRAM, DCC (Duty Cycle Corrector) should be employed. This chapter also includes the issues and requirements of DCC. Twenty papers about DLL are categorized according to their research area. In Chap. 4, general transceiver design techniques for the DRAM interface are introduced. The design issues in the transceiver arise from the imperfections in the channel characteristics. Therefore, the effects of channel are discussed, and the solutions from previous works are introduced. Also, several design techniques which are not widely adopted in other interfaces are introduced. The general concepts of the transceiver and advanced design techniques help to understand the high-bandwidth DRAM interface. Chapter 5 investigates the TSV (Thru Silicon Via) interface. TSV is a good solution to increase the bandwidth of DRAM with low power consumption. It is helpful to understand the basic configurations, architecture and issues of TSV based DRAM. One method for removing data confliction is investigated.

In the near future, further bandwidth increase is highly demanding. Therefore many readers are interested in design techniques for high-speed memory interface. This book will guide those engineers.

Seoul, Korea	Chulwoo Kim
Gyeonggi-do, Korea	Hyun-Woo Lee
Seoul, Korea	Junyoung Song
August 2013	

Contents

1	**An Introduction to High-Speed DRAM**	1
	1.1 What is DRAM?	1
	1.2 The DRAM Array	2
	1.3 DRAM Basic	4
	1.3.1 Configuration of 1 Gb DDR3 SDRAM	5
	1.3.2 Basic Operation of DDR3 SDRAM	5
	1.3.3 DRAM Core Operation	7
	1.4 Power Considerations	8
	1.4.1 Power for Activation	9
	1.4.2 Power of Standby Mode and Power Down Mode	9
	1.5 Example of DRAM	10
	References	11
2	**An I/O Line Configuration and Organization of DRAM**	13
	2.1 DRAM Core Access Speed	13
	2.1.1 DDR1	13
	2.1.2 DDR2	13
	2.1.3 DDR3	14
	2.1.4 DDR4	16
	2.2 Organization of DRAM	17
	2.2.1 Differences of DRAM Type	18
	2.2.2 Comparison of Conventional DRAM	18
	2.2.3 Comparison of Graphics DRAM	19
	2.2.4 Comparison of Mobile DRAM	20
	2.3 Application	22
	2.4 Trend of DRAM Performance	23
	References	24
3	**Clock Generation and Distribution**	25
	3.1 Why DLL Is Used For DRAM?	25
	3.2 Basic Concept of DLL	26

		3.2.1	Timing Diagram and Architecture of Delay Locked Loop . . .	27
		3.2.2	tDQSCK Skew and Locking Consideration	30
	3.3	Synchronous Mirror Delay (SMD)		32
	3.4	Register Controlled DLL		34
		3.4.1	Delay Control Method for Register Controlled DLL	34
		3.4.2	Boundary Switching Problem	34
		3.4.3	Merged Dual Coarse Delay Line.	36
		3.4.4	Delay Line Mismatch Issue	37
		3.4.5	Adaptive Bandwidth DLL in DRAM	37
	3.5	CAS Latency Controller		38
	3.6	Duty Cycle Corrector		41
	3.7	DLL Parameters for DRAM		43
	3.8	Clock Distribution of GDDR5		45
	References			47
4	**Transceiver Design**			51
	4.1	Lossy Channel		51
	4.2	Equalization Techniques		52
	4.3	Crosstalk and Skew Compensation		60
		4.3.1	Crosstalk	60
		4.3.2	Skew.	62
	4.4	Input Buffer		66
	4.5	Impedance Matching		69
	4.6	DBI and CRC		73
	References			75
5	**TSV Interface for DRAM**			77
	5.1	The Need for TSV in DRAM		77
	5.2	Die Stacking Package		77
	5.3	Configuration of DRAM and MCU/GPU via TSV		78
	5.4	Types of TSV-Based DRAM		79
	5.5	Issue of TSV-Based DRAM		81
	References			86
Index				87

Chapter 1
An Introduction to High-Speed DRAM

1.1 What is DRAM?

These days most people use at least one personal computing device, like a pc, a tablet or a smartphone. And many of them have heard about DRAM, dynamic random access memory, even though they may not know what it is. DRAM is one of the data storage components in these devices. Dynamic means it is volatile and the stored data will be erased when the device is turned off. 'Random access memory' means that we can access the data randomly, not just sequentially. In the past, people used tape as storage device. Data stored in a tape cannot be accessed randomly, only sequentially. Random access memory was developed because this sequential data access is an obstacle for high-speed operation. To increase the speed even more, synchronous DRAM, or SDRAM, was developed. 'Synchronous' means that operation is synchronized to an external clock for high-speed operation. These days, in general, DRAM means SDRAM. SDRAM can be categorized into two basic types, one is SDR and the other is DDR as shown in Fig. 1.1. SDR stands for single data rate and DDR stands for double data rate. The difference between SDR and DDR is defined by the number of input and output data per one clock period. SDR was developed first and then DDR was developed to double the data rate. There are also different types of DDR SDRAM based on how it is used. The first is DDR, which is commonly called main memory. It is mostly used in PCs and servers. Second is GDDR, which stands for graphics DDR. It is used in graphics cards and game consoles. The last is called LPDDR or mobile memory, which means low power DDR memory. It is used in mobile phones and tablets. DRAM is controlled by memory controller unit, or MCU. DRAM receives clock and command from the MCU. Data can come from the MCU for storage on DRAM or output to the MCU from the DRAM. Obviously, the DRAM needs some time for internal data processing once it receives a command from the MCU. So there are many timing specifications between the DRAM and the MCU, and CAS latency and burst length are two of the representative timing specifications. CAS latency represents the time from CAS command to first data output as shown in Fig. 1.1. It is the minimum time for reading data from the DRAM to the MCU. Burst length is the number of sequential output data per one CAS command. Due to high bandwidth requirement, burst length is becoming larger and larger [4–7].

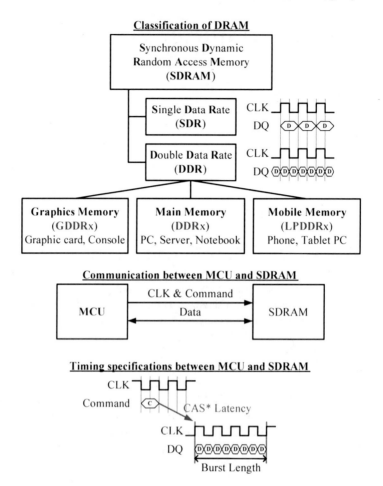

Fig. 1.1 Introduction of DRAM

1.2 The DRAM Array

Before starting a more detailed examination of high-bandwidth memory, basic knowledge of a DRAM cell array is necessary for a clear understanding of the speed limitation of DRAM core operation. However, in this book, not all kinds of circuits are included, such as array configurations, sense amplifier arrays, repair schemes, and row and column decoder elements. From the perspective area and speed, the basic cell operation with 1T1C (one transistor and one capacitor) and a sense amplifier is developed. Figure 1.2 shows the standard cell configuration. There are cell capacitors for storing data, bit line and/bit line for data sensing, word lines decoded by a row decoder, an internal column address strobe decoded by a column decoder, and a sense amplifier for amplifying data with small charge differences between a reference and data bit lines. In Fig. 1.3, the reference bit line is/bit line and data bit

1.2 The DRAM Array

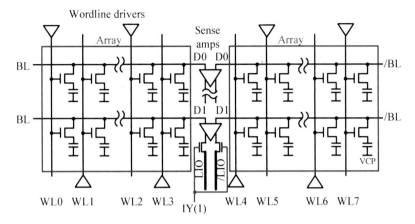

Fig. 1.2 Cell array configuration

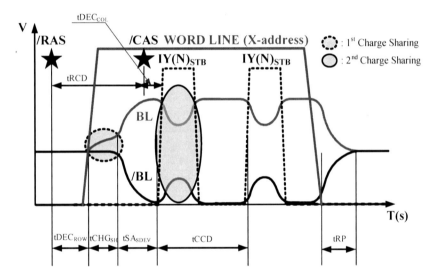

Fig. 1.3 Timing diagram

line is bit line. The RTO (meaning RESTORE) and SZ (meaning Sensing Zero) of the sense amplifiers are pre-charged to VBLP (voltage of bit line pre-charge) in the pre-charge stand-by state. The word line and internal column address strobe are in the state of logic LOW before the active command is issued. The VCP represents the voltage of the cell plate.

When the active command is decoded by the state machine and the row address is issued, the word line is activated. The word line voltage is the highest voltage of DRAM to compensate the Vt loss, because the DRAM cell configuration is based on NMOS transistor array. After the time of the word line is selected, the charge sharing occurs between the cell and the bit line. For example, if HIGH data is stored

Table 1.1 tCCD comparison

	DDR1	DDR2	DDR3	DDR4	GDDR3	GDDR4	GDDR5
tCCD (tCK)	1	2	4	4	2	4	2[a]
Pre-fetch (bits)	2	4	8	8	4	8	8

[a] tCCD of GDDR5 is 2, because CK is half rate of WCK

at the cell capacitor, the voltage of the bit line is slightly increased. However, the /bit line is not changed. Then the RTO and the SZ of the sense amplifier are developed to VCORE (Voltage for DRAM cell array) and VSS, respectively. The voltage of bit line goes to VCORE and the voltage of /bit line approaches VSS. The cell charge is restored to VCORE simultaneously. The refresh operation is just active and pre-charge operations processed by a specified time interval. The column command and column address are followed by an active command. The signal of the word line selection is level-sensitive; the word line is enabled by active command and disabled by a pre-charge command. But, the internal column address is a pulse-type. This is called a "strobe". The strobe interval is determined by the circuit designer and the operation speed. Therefore, the internal column address strobe is activated and the second charge sharing occurs. The charge of the bit line sense amplifier is transferred to local I/O and the 2nd sense amplifier to drive the global data line. The global data line utilizes a single-ended signaling, and bit lines and LIO (local I/O) lines utilize differential signaling.

To obtain data from the cell array in the pre-charge state, it takes a certain amount of time equivalent to the sum of $tDEC_{ROW}$ (row address decoding time), $tCHG_{SH}$ (charge sharing time), and tSA_{DEV} (S/A development time). Only after this amount of time, can the internal column address strobe be activated. But, after the word line is activated, another internal column address can be issued after the previous one. This is called tCCD (time of column to column delay). This time is restricted to the limitation of the sum of the 2nd charge sharing and data transferring time for global I/O lines. The second internal column address strobe can occur only after the first data is completed. tCCD is not as fast as the I/O speed. Therefore, a pre-fetch scheme is adopted to increase the core bandwidth. DDR3 has an 8-bits pre-fetch scheme. The more pre-fetch data are processed, the higher I/O bandwidth is possible. However, there is a big penalty. Due to the pre-fetch scheme, the length of burst data increases. To access another column address, it is necessary to wait to process the burst data, even if this is not welcomed by the user. Table 1.1 shows the tCCD comparisons [8–10].

1.3 DRAM Basic

It is useful to look into the operation of DRAM to understand how the high-speed DRAMs are operated. In this section, we describe the various modes of DRAM operation.

1.3.1 Configuration of 1 Gb DDR3 SDRAM

By looking at the 1 Gb (64Meg × 16) DDR3 SDRAM shown in Fig. 1.4, we explain the basic configuration of DRAM [10]. The command pins are /RAS (Row Address Strobe), /CAS (Column Address Strobe), /WE (Write Enable), /CS (Chip Selector), /CKE (Clock Enable), /RESET (Device reset), ODT (On Die Termination), and A12 for BC (Burst Chop). The ZQ register is connected to the ZQ pin for calibration of the impedance for the termination and output drivers. The address multiplexing scheme is employed to reduce the address pins. The total address pins for 1 Gb DDR3 SDRAM consist of A[11:0] for memory array and BA[2:0] for bank selection. DQ[0:15] pins are for the data interface, and DQS and DQS# are assigned to each 8-bits of DQ. Therefore, LDQ and LDQS# are for the lower 8-bits of DQ; DQ[0:7] and UDQ and UDQS# are for the upper 8-bits of DQ; DQ[8:15]. A bank is a unit of cells grouped for efficient access control and is composed of cell arrays and circuits for access control. A cell is selected by row and column addresses, which are decoded by row and column decoders. The output of row address decoder controls the word line driver, so that the selected word line will be accessed. The output of the column address decoder selects the data to be amplified by bit line sense amplifier. The selected data is amplified by read amplifier or stored by write driver. The peripheral circuit is composed of input buffers (clk, address, command input buffer), command decoder, internal voltage generator, DLL, DQ RX, DQ TX, serial to parallel and parallel to serial conversion circuits. Command from the MCU is first buffered and then goes to the command controller. The command decoder uses the external commands to generate internal control signals. Address is also received by input buffer and goes to row and column decoders. CLK is a differential signal, so CLK and CLKB are received. This CLK is used for synchronous control of command and input of DLL. The DLL controls the read drivers, which are used to control the data output. The generator makes various internal supply voltages used in the DRAM. The DRAM core has a speed limitation due to the storage cell capacitance, bit line capacitance, word line loading and so on. So the DRAM uses parallel data processing to achieve high bandwidth operation. Serial to parallel and parallel to serial circuits translate the data between inner DRAM and outer DRAM. This architecture is not identical for all DRAMs. GDDR and LPDDR are a little different. For example, an LPDDR doesn't have DLL, GDDR has more DQ RX and TX at the right side of the CMD controller and so on.

1.3.2 Basic Operation of DDR3 SDRAM

The timing diagrams of basic DRAM operation and data strobe (Read operation) are shown in Fig. 1.5. In order to access DRAM, the active command (ACT) should be issued first with row and bank addresses. Another bank active command can be issued with row and bank addresses after t_{RRD} (RAS command to RAS command delay). The total row address count is 8K (8192). One of 8K addresses is selected by an issued row address. The active command interval time is defined as t_{RRD}. After

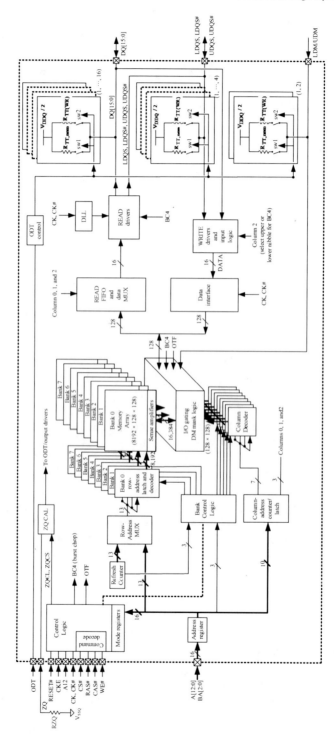

Fig. 1.4 64 Meg × 16 Functional block diagram

1.3 DRAM Basic

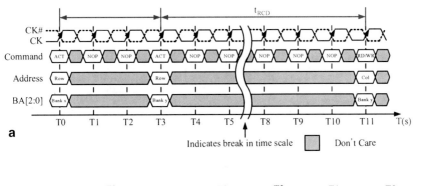

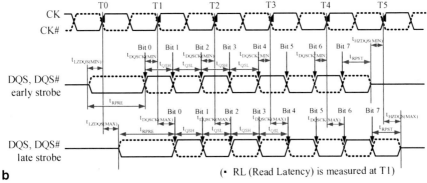

Fig. 1.5 a Timing diagram of basic DRAM operation and **b** of data strobe—read operation

the active command, a read or write (RD/WT) command can be executed with the column and bank address after t_{RCD} (RAS to CAS Delay). It is useful to look into the read operation in detail to gain insight into how high-speed DRAMs are operated. A high-speed delay locked loop (DLL) is used for the read operation. Data are output at the rising and falling edges of CK, corresponding to a double data rate (DDR). The ideal position of the rising edge of DQS should be aligned at the rising edge of CK. However, DQS is shifted to the left or right of CK due to mismatch between the real and replica paths of DLL. This will be developed in Chap. 3.

1.3.3 DRAM Core Operation

The four basic operations of DRAM are ACTIVE, PRECHARGE, WRITE, and READ. ACTIVE and PRECHARGE means the selected WL(word line) is enabled and disabled, respectively. WRITE means storing the data to memory cells and READ means reading the data from memory cells. In Fig. 1.6, we can easily understand WRITE and READ operations. The red line illustrates the dataflow of a write operation. DQ RX receives the serial data from the MCU. The received data is parallelized by the serial to parallel circuit and transferred to the bank using GIOs

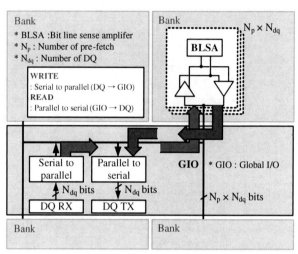

Fig. 1.6 Data flow of DRAM

(Global I/O lines). The GIOs are very long channels and are connected with all the banks. The transferred data is stored in the cell through WRITE driver and bit line sense amplifier (BLSA). The blue line shows the dataflow of a read operation. The stored data in the memory cell is amplified by the BLSA and the main amplifier which is in the bank. And then, the data is fed to the parallel to serial circuit by the GIOs. The transferred data is aligned and serialized in the parallel to serial circuit. Finally, the serialized data is output synchronously with the DCLK, which is a somewhat delayed clock by the DLL. The number of GIO channels is the pre-fetch number multiplied by the number of DQs. Parallel GIO channels are the main technique for achieving high data rate memory. But the number of GIO channels can be a bottleneck as well, because the more GIO channels, the larger channel area, which limits the speed increase.

1.4 Power Considerations

Power consumption is one of the biggest concerns to increase performance for mobile devices and computing systems. Due to limited battery life and instant-on conditions, low power requirements are considered more significantly. Because of higher data rates and a number of memory channels, it is very hard to reduce the power consumption. The memory subsystem is the second biggest power drain after the processor. Low power and high performance have an inevitable trade-off relationship in currently developed systems. The higher-performance systems such as high-definition graphics cards dissipate power to transfer data with the number of memory channels for increasing the data bandwidth. There are several approaches to save power. Dynamic voltage/frequency scaling (DVFS) is one good solution to extend battery life. It is helpful for readers to examine the two critical memory operating modes

1.4 Power Considerations

Table 1.2 IDD description and current of 2G DDR3-1600

	Description	Current (2G DDR3-1600)
IDD0	Operating one bank active pre-charge current	45 mA
IDD1	Operating one bank active read pre-charge current	55 mA
IDD2N	Pre-charge standby current (or pre-charge non-power-down current)	25 mA
IDD2P	Pre-charge power-down current	12 mA
IDD3N	Active power-down current (or Active non-power-down current)	30 mA
IDD3P	Active power-down current	17 mA
IDD4R	Operating burst read current	105 mA
IDD4W	Operating burst write current	95 mA
IDD5	Burst refresh current	120 mA
IDD6	Self-refresh current	12 mA
IDD7	Operating bank interleave read current	185 mA

for understanding memory power consumption. Also, IDD description and current of 2G DDR3-1600 is summarized in Table 1.2.

1.4.1 Power for Activation

In order to access a DRAM, the active command should be issued with a row address. When the active command is issued, many sense amplifiers of the cell array are developed for amplifying the data of the cell capacitors. Though it depends on the DRAM configuration, at least 512 sense amplifiers are operated simultaneously. Most of the power of the cell array is drained by the sense amplifiers. After the activation command, a READ or WRITE command can be executed. The READ command needs to drive the global I/O lines to transfer data from the cell array to pipe-registers, and the DLL is awaked to output data at the rising edge of the external clock. In general, DRAM has several banks. These banks can be operated in interleave mode to increase the number of consecutive data. IDD7 operates a bank interleave read current. IDD7 is the largest power-consuming mode. DRAM datasheets provide the IDD power according to operating modes. In the case of a single unbuffered 1 GB DDR3 module, about 5 W and an additional 2 W of I/O power are dissipated. Therefore, the dual-channel memory system can easily exceed 10 W. If quad processor is adopted in a system, the memory can easily dissipate the power of 100 W. For future memory power, new solutions or circuit innovations should be studied and developed. As the memory bandwidth increases more, the supply voltage will be lowered to reduce the power consumption. Therefore, low-voltage circuits have also been needed.

1.4.2 Power of Standby Mode and Power Down Mode

When a DRAM is not accessed, it is possible for it to stay in standby or power-down mode. In standby mode, memory power dissipation decreases because the

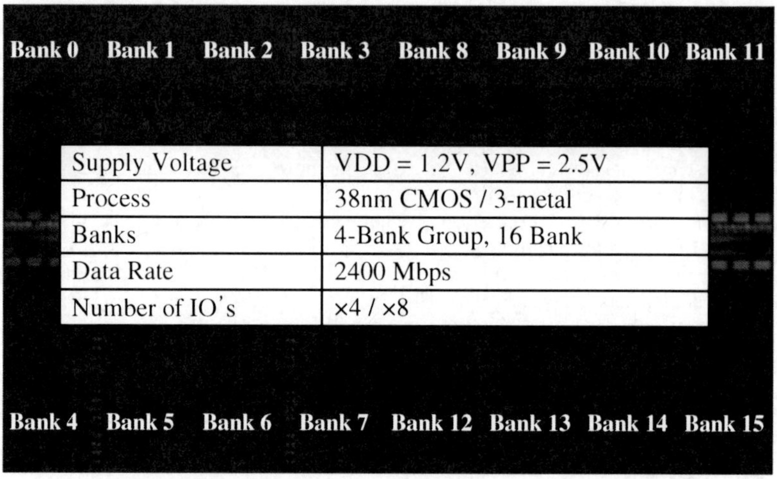
Bank 0
Bank 1
Bank 2
Bank 3
Bank 8
Bank 9
Bank 10
Bank 11
Supply Voltage
VDD = 1.2V, VPP = 2.5V
Process
38nm CMOS / 3-metal
Banks
4-Bank Group, 16 Bank
Data Rate
2400 Mbps
Number of IO's
×4 / ×8
Bank 4
Bank 5
Bank 6
Bank 7
Bank 12
Bank 13
Bank 14
Bank 15

Fig. 1.7 Real die photo of DDR4 SDRAM

DRAM core is not activated. However, because DRAM should respond directly to an MCU's command from standby mode to active, peripheral circuits must be awakened. Therefore, a number of Rx without DQs and DLL are awakened. Though the power consumption is reduced substantially, power dissipation does not approach zero. Otherwise, when a DRAM is in power-down mode, all Rx are turned off without the CKE buffer, and the DLL enters a frozen state. DRAM only needs to sustain the internal voltage, especially VPP and VBB, which are generated by a charge pump. However, the power dissipation in power-down mode is less than 5 % of its peak power. Many computing systems are not utilized under peak workload at all times. Therefore, the power dissipation in standby and power-down modes becomes a crucial point for total memory power. For example, if a workstation that has multiple modules is in active mode with only one module, the modules of remaining ranks are in standby mode. Under peak workloads, standby power of DDR3 employs a high-capacity server that accounts for up to 60 % of the total memory power. For mobile PCs which need to extend the battery life, power-down mode is frequently utilized to save standby power. If mobile PCs enter a power-saving mode that uses the self-refresh mode of DRAM, the power consumption of memory systems is reduced to about 2 % of its peak power. However, in order to exit self-refresh mode, it is necessary to wait for 500 clock cycles to relock a DLL.

1.5 Example of DRAM

Figure 1.7 is a DDR4 die photo, which was introduced in ISSCC 2013 [11]. The data is stored in memory cells, each of which consists of one capacitor and one transistor. It has a huge number of memory cells, which is grouped by several banks for efficient access control. The memory control circuits are located in the center of the chip and it occupies a very small area. I would like you to know the large size of the bank and the overall DRAM structure from this picture as we introduced before.

References

1. R. Rho et al., "A 75 nm 7 Gb/s/pin 1 Gb GDDR5 graphics memory device with bandwidth improvement techniques," *IEEE J. Solid-State Circuits*, vol. 45, no. 1, pp. 120–133, Jan. 2010.
2. T.-Y. Oh et al., "A 7 Gb/s/pin 1 Gbit GDDR5 SDRAM with 2.5 ns bank to bank active time and no bank group restriction," *IEEE J. Solid-State Circuits*, vol. 46, no. 1, pp. 120–133, Jan. 2011.
3. D. Shin et al., "Wide-range fast-lock duty-cycle corrector with offset-tolerant duty-cycle detection scheme for 54 nm 7 Gb/s GDDR5 DRAM interface" *IEEE Symp. on Very Large Scale Integr. Circuits Dig. Tech. Papers*, 2009, pp. 138–139.
4. JEDEC, JESD79F.
5. JEDEC, JESD79-2F.
6. JEDEC, JESD79-3F.
7. JEDEC, JESD79-4.
8. http://www.hynix.com/inc/pdfDownload.jsp?path=/datasheet/Timing_Device/DDR_Device Operation.pdf
9. http://www.hynix.com/inc/pdfDownload.jsp?path=/datasheet/Timing_Device/DDR2_device_operation_timing_diagram_Rev.0.1.pdf
10. http://www.hynix.com/inc/pdfDownload.jsp?path=/datasheet/Timing_Device/Consumer DDR3 Device Operation
11. K. Koo et al., "A 1.2 V 38 nm 2.4 Gb/s/pin 2 Gb DDR4 SDRAM with bank group and × 4 half-page architecture", in *ISSCC Dig. Tech. Papers*, pp. 40–41, Feb. 2012.

Chapter 2
An I/O Line Configuration and Organization of DRAM

2.1 DRAM Core Access Speed

For state-of-the-art DRAM, the core operating speed is around 200 Mb/s. However, data is transferred by 7 Gb/s/pin for GDDR5 [1–3]. This is possible because pre-fetch scheme is employed. In this session, we explain the pre-fetch scheme and global IO configuration for understanding speed limits restricted by DRAM core operation.

2.1.1 DDR1

Figure 2.1 shows a gapless read operation in DDR1 [4]. Burst length of DDR1 is two. It means two data bits are sequentially output from the memory for one read operation. CAS to CAS delay of DDR1 is one clock period. This means the minimum gap between CAS to CAS commands is one clock period. Therefore, DDR1 outputs two data bits in one clock read operation. As mentioned earlier, DRAM core has a speed limitation due to the various capacitive loads and sensing operation of bit line sense amplifier. So, to get around this problem, DDR1 reads two data bits at the same time. The two data bits that are read out from cell array simultaneously is called 'pre-fetched data' As a result, the number of GIO (Global I/O) can be calculated by the required burst length multiplied by the number of DQ. As explained earlier, the number of GIO channels is the number of pre-fetch bits multiplied by the number of DQ as shown in Fig. 2.2. And the core operation speed is half of the output data rate for DDR1. For example, if the output data rate is 400 Mb/s, the internal data rate is 200 Mb/s.

2.1.2 DDR2

This timing diagram of Fig. 2.3 shows a gapless read operation in DDR2 [5]. The burst length of DDR2 is four. It means four consecutive data bits are output from the memory for one read operation. CAS to CAS delay of DDR2 is two clock cycles. The tCCD is doubled compared with DDR1 in terms of clock cycles. But DDR2

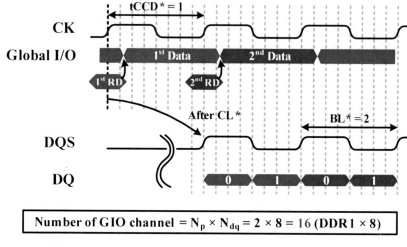

Number of GIO channel = $N_p \times N_{dq} = 2 \times 8 = 16$ (DDR1 $\times$ 8)

* tCCD : CAS to CAS delay * BL : Burst length
* CL   : CAS latency

Fig. 2.1 Timing diagram of read operation for DDR1

Fig. 2.2 Configuration of pre-fetch and GIOs connection for DDR1

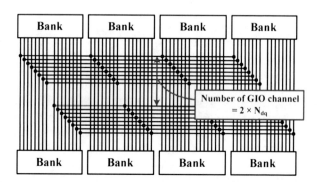

clock speed is twice as fast as DDR1. DDR2 reads four data bits from cell array at the same time. As a result, the number of GIO is the necessary burst length multiplied by the number of DQ. DDR2 needs four data reading at the same time. So DDR2 needs four GIO channels for each DQ. And the core operation speed is quarter of the output data rate. So if the output data rate is 800 Mbps, the internal data rate is 200 Mbps. The internal data rate of DDR2 is the same as DDR1. The configuration of pre-fetch and GIOs connections for DDR2 is shown in Fig. 2.4.

2.1.3 DDR3

This timing diagram of Fig. 2.5 represents a gapless read operation in DDR3 [6]. The burst length of DDR3 is eight. That means eight consecutive data bits come

YBP Library Services

KIM, CHULWOO.

HIGH-BANDWIDTH MEMORY INTERFACE.

                                         Paper   88 P.
NEW YORK: SPRINGER, 2014
SER: SPRINGERBRIEFS IN ELECTRICAL AND COMPUTER ENGINEERING.
AUTH: KOREA UNIVERSITY.

LCCN 2013950761
  ISBN 3319023802   **Library PO#** SLIP ORDERS

		List	54.99	USD
6207 UNIV OF TEXAS/SAN ANTONIO		**Disc**	17.0%	
App. Date 4/09/14 CSC.APR	6108-09	**Net**	45.64	USD

SUBJ: COMPUTER STORAGE DEVICES.

CLASS TK7895   DEWEY# 621.397   LEVEL ADV-AC

YBP Library Services

KIM, CHULWOO.

HIGH-BANDWIDTH MEMORY INTERFACE.

                                         Paper   88 P.
NEW YORK: SPRINGER, 2014
SER: SPRINGERBRIEFS IN ELECTRICAL AND COMPUTER ENGINEERING.
AUTH: KOREA UNIVERSITY.

  LCCN 2013950761
    ISBN 3319023802   **Library PO#** SLIP ORDERS

		List	54.99	USD
6207 UNIV OF TEXAS/SAN ANTONIO		**Disc**	17.0%	
App. Date 4/09/14 CSC.APR	6108-09	**Net**	45.64	USD

SUBJ: COMPUTER STORAGE DEVICES.

CLASS TK7895   DEWEY# 621.397   LEVEL ADV-AC

2.1 DRAM Core Access Speed

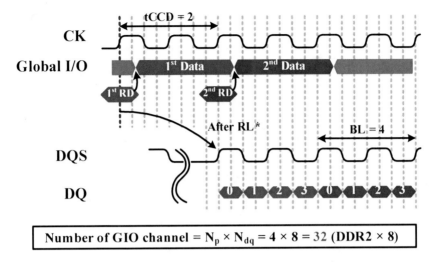

* RL : READ latency

Fig. 2.3 Timing diagram of read operation for DDR2

Fig. 2.4 Configuration of pre-fetch and GIOs connection for DDR2

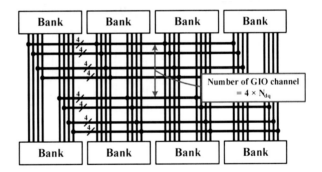

from the memory for one read operation. CAS to CAS delay of DDR3 is four clock cycles. The tCCD is doubled compared to DDR2 in terms of clock cycles. DDR3 clock speed is twice as fast as DDR2. Even though the eight bits consecutive data are not required, DDR3 reads eight data at the same time. In order to chop the burst data, DDR3 supports the burst chop function. So, you can clearly see the trend. To increase the data rate, the number of pre-fetched bits is increased. The configuration of pre-fetch and GIOs connections for DDR3 is shown in Fig. 2.6.

Unfortunately, the number of GIO channels is increased by the same factor as well. This increase in GIO channels causes the channel area to increase. It is a major disadvantage of the pre-fetched scheme.

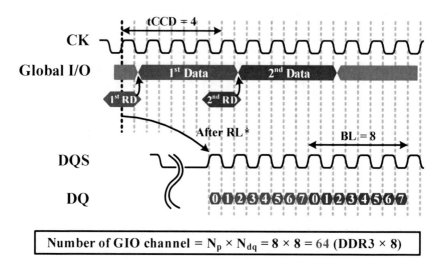

Fig. 2.5 Timing diagram of read operation for DDR3

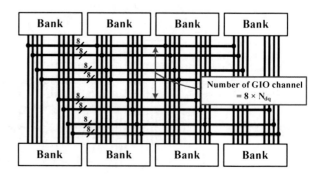

Fig. 2.6 Configuration of pre-fetch and GIOs connection for DDR3

2.1.4 DDR4

Finally, Fig. 2.7 illustrates a gapless read operation in DDR4 [7–8]. If we follow the previous trend in burst length, the burst length of DDR4 should be 16. But too large of a burst length is inefficient for data processing. So, the burst length is fixed to eight, the same as DDR3. Due to the core speed limitation, you can see a one clock period gap between gapless read operations when tCCD of DDR4 is five. To get rid of this wasted time, a bank group architecture is adopted in DDR4 [9]. The bank group architecture is a very simple idea. The banks that are in the same bank group share the GIO channel. If the interleaving read is applied and each access bank is not in the same bank group, the output data can be transferred to each GIO channel in time. So the output data bits do not collide and the gap of interleaving read can be reduced so that tCCD is four. This bank group interleaving access helps to get rid of wasted time. Therefore, DDR4 has two different CAS to CAS delays, tCCD_S (tCCD Short) and tCCD_L (tCCD Long). tCCD_S is the gap of interleaving read command from different bank groups. tCCD_L is the gap of interleaving read command from the

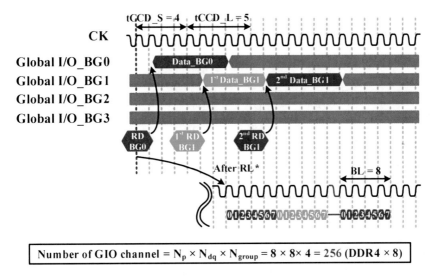

$$\text{Number of GIO channel} = N_p \times N_{dq} \times N_{group} = 8 \times 8 \times 4 = 256 \text{ (DDR4} \times 8)$$

Fig. 2.7 Timing diagram of read operation for DDR4

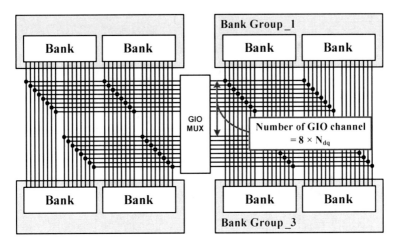

Fig. 2.8 Configuration of pre-fetch and GIOs connection for DDR4

same bank group. Figure 2.8 is a pre-fetch and bank grouping diagram of DDR4. There are no rules as far as how to group the banks. Each channel of a bank group is connected to GIO_MUX, and the GIO_MUX selects the suitable channel.

2.2 Organization of DRAM

In general, it is possible to categorize the DRAM based on their types and applications. DRAM has different features for different applications. We briefly look into their unique features according to their types.

Table 2.1 Differences of DDRx, GDDRx, and LPDDRx

	GDDRx	DDRx	LPDDRx
Archi-tecture	Bank | Bank PAD Bank | Bank	Bank | Bank PAD Bank | Bank	PAD Bank | Bank Bank | Bank PAD
Socket	On board	DIMM	MCP*/PoP*/SiP*
IO	×16 / ×32	×4 / ×8	×16 / ×32
Application	Graphic card	PC/Server	Phone/Consume
Application	• Single uni-directional WDQS, RDQS • VDDQ termination • CRC, DBI • ABI		• No DLL • DPD * • PASR * • TCSR *

* MCP: Multi chip package
* PoP : Package on package
* SiP : System in package
* DPD : Deep power down
* PASR: Partial array self refresh
* TCSR: Temperature compensated self refresh

2.2.1 Differences of DRAM Type

Table 2.1 shows the differences in the three types of DDR. The pad locations of DDR and GDDR are in the center of the chip. On the other hand, LPDDR has pads at edge of the chip due to the different package type. It is good to make multi chip package. LPDDR is normally not packaged alone. So it has an edge pad layout for multi-chip package. Each type of memory has unique features. GDDR is optimized for high speed and large bandwidth. So GDDR has x16 or x32 IO and cyclic redundancy check (CRC) feature. LPDDR is optimized for low power consumption. So LPDDR does not use DLL and supports various functions for lower power consumption such as deep power down mode, partial self-refresh and temperature compensated self-refresh. Recently developed DDR4 has gear down mode, CRC and DBI features for high-speed operation and TCSR for low power. Hence, the differences among DDR, GDDR and LPDDR are becoming less and less. DDR4 borrow some features of GDDR5 such as VDDQ termination, DBI and so on.

2.2.2 Comparison of Conventional DRAM

Table 2.2 shows the differences in DDR1, DDR2, DDR3, and DDR4. Supply voltage is scaled down for low power consumption. Starting from DDR2, OCD and ODT are utilized. OCD is off-chip driver impedance adjustment for the voltage output level for DQ and DQS. So it helps to maximize the valid data window amplitude. ODT is on-die termination to reduce the signal reflection. Starting from DDR3, dynamic ODT,

2.2 Organization of DRAM

Table 2.2 DDR comparison

	DDR1	DDR2	DDR3	DDR4
VDD [V]	2.5	1.8	1.5	1.2
Data rate [bps/pin]	200 ~ 400 M	400 ~ 800 M	800 M ~ 2.1 G	1.6 ~ 3.2 G
STROBE	Single DQS	Differential DQS, DQSB		
Pre-fetch	2 bit	4 bit	8 bit	8 bit
Interface	SSTL_2	SSTL_2	SSTL_15	POD-12
New feature		ODT	Write levelling	CA parity
		OCD calibration	Dynamic ODT	Bank grouping
			ZQ calibration	Gear down
				DBI, CRC
				CAL
				PDA
				FGREF
				TCAR

DBI Data bus inversion, *CAL* Command address latency, *FGREF* Fine ganularity, *CRC* Cyclic redundancy check, *PDA* Per DRAM adressability, *TCAR* Temparature controlled array refresh

ZQ calibration and write leveling are applied. Dynamic ODT mode is for changing the termination strength of the memory without issuing an MRS command. MRS stands for mode register set and is used as a command set to control the memory mode. The dynamic ODT mode supplies two RTT values which are RTT_Nom and RTT_WR. ZQ calibration is used to calibrate DRAM Ron and ODT values. DDR3 memory module adopted a fly-by topology for better signal integrity. Fly-by topology has the benefits of reducing the number of stubs and their lengths, but it creates flight time skew between the clock and DQS at each DRAM. Therefore, DDR3 supports write leveling to reduce this skew of DQ. DDR4 has dual error detection schemes. The first one is cyclic redundancy check for DQ pins. And the other is command address parity for command and address pins. The CRC scheme is used with data-bus inversion scheme. Data bus inversion reduces the number of switching at DQs to reduce the I/O current. This also improves signal integrity because it reduces the simultaneous switching output (SSO) noise.

2.2.3 Comparison of Graphics DRAM

Now let us examine the graphics memories from Table 2.3. GDDR5 has many new features. For example, it has a double data rate address scheme. At the rising edge of CK, command and half of the address are issued. At the rising edge of CKB (which is the falling edge of CK), remaining half of the address are issued. GDDR5 has differential data clock inputs, called WCK and WCKB. These are called write clocks. Write data are referenced to both edges of a free-running differential forwarded clock (WCK, WCKB) which replaces the pulsed strobes used in conventional graphic memories such as GDDR3 as it called WDQS. It also includes address bus inversion logic and ABI (Address bit inversion) pin. This means that the number of pull down switching pins is reduced to less than four or five out of nine to lower the power

Table 2.3 GDDR comparison

	GDDR1	GDDR2	GDDR3	GDDR4	GDDR5
VDD [V]	2.5	1.8	1.5	1.5	1.5/1.35
Data rate [bps/pin]	300 ~ 900 M	800 M ~ 1 G	700 M ~ 2.6 G	2.0 ~ 3.0 G	3.6 ~ 7.0 G
STROBE	Single DQS	Differential Bi-directional DQS, DQSB	Single unidirectional WDQS, RDQS		
Pre-fetch	2 bit	4 bit	4 bit	8 bit	8 bit
Interface	SSTL_2	SSTL_2	POD-18	POD-15	POD-15
New feature		ODT OCD calibration	ZQ	DBI Parity (opt)	Bank grouping No DLL PLL (option) RDQS (option) WCK, WCKB CRC ABI

DQS DQ strobe signal, DQ is data I/O pin, *ODT* on die termination, *OCD* Off chip driver, *ABI* address bus inversion

consumption. If the ABI pin is high, command and address pins are not inverted. If the ABI pin is low, command and address pins are inverted. Bank grouping is a solution for securing the core timing margin for high-speed operation. POD (Pseudo Open Drain) is employed in GDDR memory. VDDQ termination is adopted. Only the low data consumes the energy from the channel. The interface scheme is illustrated in Fig. 2.9. In SSTL, VREF is 0.5 × VDDQ. Therefore, the signal center point is a half of VDDQ. In order to transfer data via the channel, HIGH and LOW data consume power. Otherwise, the VDDQ termination type of GDDR5 only consumes power to drive LOW data to the channel. HIGH data stays in VDDQ. GDDR3, 4, 5 and DDR4 employ a POD (Pseudo Open Drain) type, which is good at high-speed operation with low power. If the DBI function is enabled, the energy consumption is more reduced. Because LOW data only consumes energy, if the number of LOW data is larger than HIGH data, then data polarity is inverted and signal DBI is activated.

2.2.4 Comparison of Mobile DRAM

In Table 2.4, we compare low power DRAMs. Data rate of LPDDR3 can reach up to 1.6 Gb/s/pin. The supply voltage of LPDDR2 and 3 are reduced to 1.2 V from 1.8 V. Here, DQS_T and DQS_C represent a pair of data strobes. DQS_C is the complement of DQS_T. Data strobe of LPDDR3 is bi-directional and differential. DQS_T is edge-aligned to read data and centered with write data. The differential DQS and ODT is happened in mobile DRAM even though they require additional power consumption to increase the operating speed. However, the ODT is not welcome to system engineers due to the high power consumption even though the speed performance is increased with the help of the ODT. The DLL is not used in LPDDR

2.2 Organization of DRAM

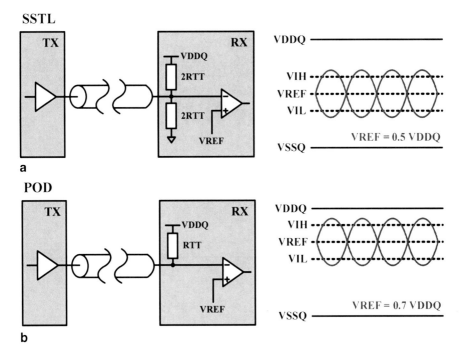

Fig. 2.9 Interface comparison between **a** SSTL for DDR2/3 and **b** POD for GDDRx

Table 2.4 LPDDR comparison

	LPDDR1	LPDDR2	LPDDR3
VDD [V]	1.8	1.2	1.2
Data rate[bps/pin]	200~400M	200 M~1.066 G	333 M~1.6 G
STROBE	DQS	DQS_T, DQS_C	DQS_T, DQS_C
Pre-Fetch	2 bit	4 bit	8 bit
Interface	SSTL_18	HSUL_12	HSUL_12
New feature	No DLL	No DLL	No DLL
		CA pin	ODT (high tapped termination)

SSTL Stub series terminated logic, *HSUL* High spped un-terminated logic

memory in order to reduce the power consumption. The typical features of low-power DRAM are as follows:

- Low voltage (additional PAD for VPP)
- Low I/O capacitance
- Un-terminated I/Os (HSUL)
- Typically X16 or X32 data width per die
- Usually contained in multi-die packages
- Fast low-power entry and exit
- Very low standby (self-fresh) power (with long refresh rate)

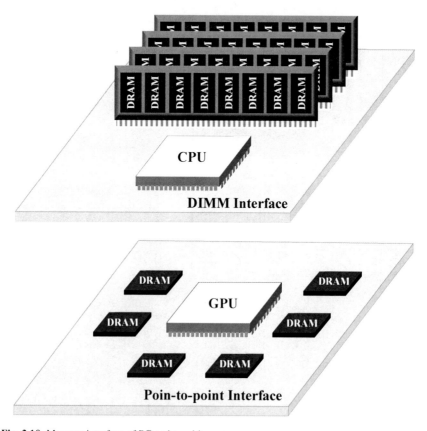

Fig. 2.10 Memory interface of PC and graphics memory

Temperature compensated self-refresh mode/ Partial array self-refresh
Deep power down mode
No DLL/ Low speed (up to LPDDR2, but high speed requirements increase)

2.3 Application

As shown in Fig. 2.10 DRAM has many single-ended data channels. A DDR memory module consists of multi rank and multi DIMM. This causes impedance discontinuities due to many stubs, connectors, vias and so on. Therefore, the data channel of DRAM is vulnerable to noise coupling and inter symbol interference, ISI. DQ drivers or receivers usually operate all at the same time. So simultaneous switching output noise (SSON) is a problem for high quality SI (Signal integrity). Otherwise, graphics DRAMs are configured with point-to-point connection. Therefore SI is better than the one of DIMM channel. Therefore, it is harder to increase the operating speed for DIMM channel than a graphics point-to-point channel. The last issue of

2.4 Trend of DRAM Performance

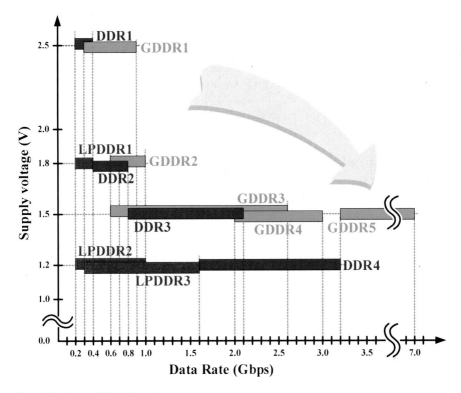

Fig. 2.11 Trend of DRAM performance

high-bandwidth DRAM design is that the transistor performance of DRAM process is very poor compared to that of logic process. Therefore, high-speed graphics DRAM such as GDDR5 sometimes needs special transistors that have large output impedance and lower threshold voltage for critical path circuit design.

2.4 Trend of DRAM Performance

Figure 2.11 shows the trend of each DRAM type. From 2.5 V to 1.2 V and from the 0.2 Gb/s/pin to 7.0 Gb/p/s, DRAM has developed rapidly. Several years ago, the speed performance was the top design priority. However, in the mobile-era, low-power consumption is the key design parameter. Some DDR2 which is configured X16 data-width can be used together for graphics applications. That is gDDR2. GDDR2 which is configured X32 data-width is not sold much in the market. Although all types of DRAMs are reaching their limits in supply voltage, the demand for high-bandwidth memory is keep on increasing. Sometimes, the supply voltage for graphics DRAM is raised to boost the speed up, which is called over-clocking. When you look at each type of DRAM as a group, each DRAM has its own special features. For example, Graphics memory was developed because of the need for high band width memory for processing mass graphics data. So graphics memory has 16 or 32 I/Os which is different from main memory (8 I/Os). 7 Gb/p/s GDDR5 is the fastest DRAM up to now. Mobile memory is specialized memory for mobile applications, which needs low power consumption. So it doesn't use DLL and has a variety of power down modes. As such, each type of DRAM had its own unique characteristics for optimum performance in its intended applications. But, these days, as technology is maturing, the differences among each type of DRAMs is becoming smaller and smaller. Design techniques of DRAM for high speed and low power are reaching saturation. Some have said that all kinds of memories will be unified into one type. However, this will take much time because the AP or MCU will need various types of memories for a while according to their applications.

References

1. R. Rho et al., "A 75 nm 7 Gb/s/pin 1 Gb GDDR5 graphics memory device with bandwidth improvement techniques," *IEEE J. Solid-State Circuits*, vol. 45, no. 1, pp. 120–133, Jan. 2010.
2. T.-Y. Oh et al., "A 7 Gb/s/pin 1 Gbit GDDR5 SDRAM with 2.5 ns bank to bank active time and no bank group restriction," *IEEE J. Solid-State Circuits*, vol. 46, no. 1, pp. 120–133, Jan. 2011.
3. D. Shin et al., "Wide-range fast-lock duty-cycle corrector with offset-tolerant duty-cycle detection scheme for 54 nm 7 Gb/s GDDR5 DRAM interface" *IEEE Symp. on Very Large Scale Integr. Circuits Dig. Tech. Papers*, 2009, pp. 138–139.
4. JEDEC, JESD79F.
5. JEDEC, JESD79-2F.
6. JEDEC, JESD79-3F.
7. JEDEC, JESD79-4.
8. T.-Y. Oh et al., "A 7 Gb/s/pin GDDR5 SDRAM with 2.5 ns bank-to-bank active time and no bank-group restriction", in *ISSCC Dig. Tech. Papers*, pp. 434–435, Feb. 2010.
9. K. Koo et al., "A 1.2 V 38 nm 2.4 Gb/s/pin 2 Gb DDR4 SDRAM with bank group and × 4 half-page architecture", in *ISSCC Dig. Tech. Papers*, pp. 40–41, Feb. 2012.

Chapter 3
Clock Generation and Distribution

DLL is a popular clock generator in DRAM and duty cycle corrector is an essential block for DDR DRAM. Low skew, low power clock distribution method has become an important issue in high-bandwidth memories.

3.1 Why DLL Is Used For DRAM?

In the single-data-rate SDRAM, it is hard for the operating speed to exceed around 100 MHz [1]. In order to increase the data rate, a double-data-rate scheme is employed. However, one of the important hurdles is that the valid data window varies according to the PVT variations. In Fig. 3.1, the valid data window is reduced with increasing operating speed and PVT variations. It is harder for the double-data-rate scheme to obtain a valid data window than the single-data-rate one. The MCU (memory control unit) cannot have a valid sampling position. Therefore, in order to transfer data with higher speed, the timing variation according to PVT variations should be reduced.

Figure 3.2 shows DRAM without a DLL. Data is transferred through an output logic circuit and output driver by the delayed system clock. An internal delay of the system clock occurs with the internal clock path (tD_1). Also, the output data is also delayed by the output logic and output driver (tD_2). tD_1 and tD_2 are varied according to PVT variations. Therefore, as shown in Fig. 3.1 the timing variation caused by PVT variations occurs through the paths of tD_1 and tD_2.

Therefore, to increase the timing margin, the output timing of DATA should not be varied according to PVT variations. A DLL helps to reduce the timing variation caused by PVT variations. In DRAM with a DLL [2], data is aligned with the rising edge of the clock. It is not varied by PVT variations. Therefore, the DLL can remove the timing variations. In addition, the DLL makes a larger valid data window and increases the timing margins as shown in Fig. 3.3. In general, DDR SDRAM has a DLL from DDR1 to DDR4 and from GDDR1 to GDDR4. GDDR5 and LPDDR SDRAM do not have a DLL. In the case of GDDR5, the MCU has a clock and data recovery (CDR) to recover data without a DLL. LPDDR SDRAM focuses on low power consumption at the expense of low-speed operation.

Fig. 3.1 High speed timing budget according to PVT without DLL

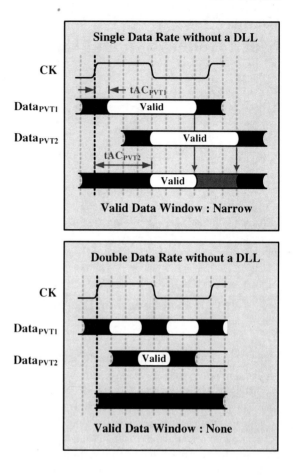

Figure 3.4 shows the DRAM that includes a DLL. From the system clock, DLL generates DLLCLK, which is called a negative delayed clock. In the timing domain, DLLCLK leads the system clock by an amount equivalent to the data output timing (tD_2). The timing of tD_{REP} is a copy of the data output delay (tD_2) and input clock's delay (tD_1). The basic operation of the DLL is explained in detail in Session 3.2.

3.2 Basic Concept of DLL

The main goal of the DLL is the generation of a clock that compensates for the voltage, temperature, and process variations. The DLL has a feedback loop to compare the external clock and DLLCLK for compensating for the temperature and voltage changes continuously. Without the DLL in high-speed DRAM, the output data is varied according to the PVT variations, because the output path consists of many logic gates. The output delay variance is dependent on the output data path. Therefore, to remove the output delay variance, the output delay should be known. If one

3.2 Basic Concept of DLL

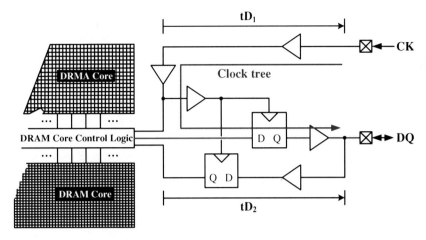

Fig. 3.2 Simplified DRAM architecture without DLL

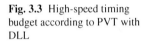

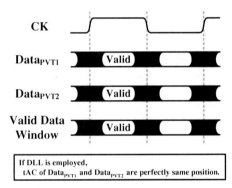

Fig. 3.3 High-speed timing budget according to PVT with DLL

circuit can make a special clock which leads the system clock by the output delay, then data can appear at the same position as the system clock. This is a main goal of the DLL. By using the replica delay that mimics the output delay, the DLL generates the negative delayed clock. We can easily copy the delay of output path by modeling the output logic path. Finally, by the help of the DLL, DRAM output data is always aligned with the rising edge of the system clock.

3.2.1 Timing Diagram and Architecture of Delay Locked Loop

DLL has been adopted in synchronous DRAM to synchronize the external input clock and the output data. From the Fig. 3.5, Data output path of DRAM is shown in gray and the white rectangle represents the DLL. The DLL consists of variable delay line, phase detector, controller and replica delay. Phase detector compares the phase of input clock and the output of the delay line noted as FB_CLK. Based on the PD

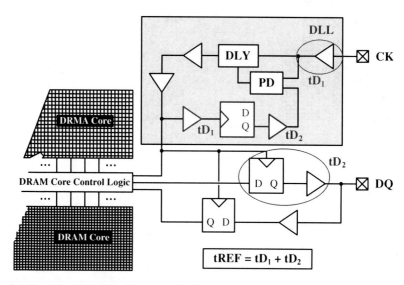

Fig. 3.4 Simplified DRAM architecture with DLL

comparison result, the controller adjusts the number of activated delay cells to lock the DLL. DLLs used in DRAM include a replica delay block to compensate for the input and output buffer delays of DRAM which are tD_1 and tD_2, respectively. They have the same timing as explained earlier. To understand the DLL's operation, it is helpful to examine the timing diagram of the DLL under the locked condition.

The timing of tD_1 is the delay length of the input path, and the timing of tD_2 is the delay length of the output path and timing of tD_{DLY} is the variable delay line length. The timing of tD_{REP} is the delay length of the replica circuit that is determined by the sum of the tD_1 and tD_2 paths. The total delay is N times tCK under the locked condition. (N is integer.) Under the locked condition of the DLL, the DRAM uses a DLL clock in pipe-latches to output the signals DQSs and DQs (output logics). The main role of the DLL is to compensate for the internal delay in order to align the data output time with the rising edge of the external CLK in all PVT conditions. The timings of tD_1, tD_2, and tD_{REP} are asynchronous delays and vary according to the PVT variations. Under the locked condition, the following equations are satisfied:

$$tCK \times N = tD_{DLY} + tD_{REP} \tag{3.1}$$

$$tD_{REP} \approx tD1 + tD2 \tag{3.2}$$

The DLL updates the length of the variable delay line (tD_{DLY}) in order to satisfy Eq. (3.1) at a given operating frequency and supply voltage. However, due to the mismatch between tD_{REP} and the sum of tD_1 and tD_2, Eq. (3.1) should be re-expressed as:

$$tCK \times N = tD_{DLY} + tD1 + tD2 + \gamma \tag{3.3}$$

$$\gamma = tD_{REP} - (tD1 + tD2), \tag{3.4}$$

3.2 Basic Concept of DLL

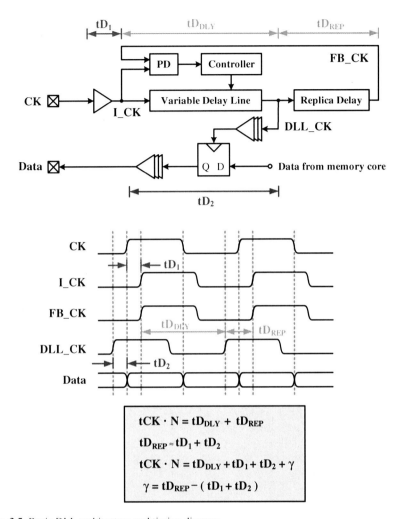

Fig. 3.5 Basic DLL architecture and timing diagram

where the parameter γ is the mismatch parameter between tD_{REP} and the sum of tD_1 and tD_2. γ varies according to the PVT variations. From Fig. 3.5, the total delay is N times tCK under locked condition, and as you can see in Fig. 3.5, I_CK is delayed by tD_1 from the CK. When the DLL is locked, phases of I_CK and FB_CLK are aligned. The phase difference between FB_CLK and DLL_CLK is $tD_1 + tD_2$. Therefore, Data comes out after tD_2 delay from DLL_CLK which means that the output data is synchronized to the external input clock. After reaching the locked condition, the DLL keeps the locked condition by continuously compensating for the changes of replica delay that copy the circuits of the input buffer and output logics. Therefore, output timing delay changes caused by PVT variations are released by the DLL. However, it is not easy to match the replica delay to $tD_1 + tD_2$. The reasons are; first, although the input buffer receives a differential, small swing signal, the

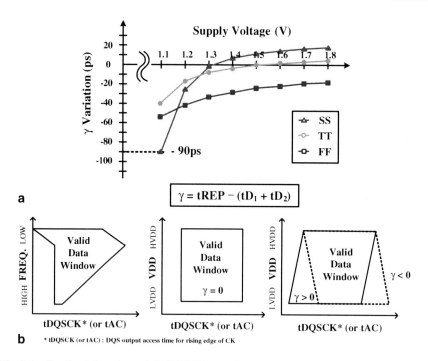

Fig. 3.6 a Replica delay mismatch b tDQSCK (or tAC) skew

replica delay block receives a single-ended full swing signal in on-chip environment. Second, the DOUT buffer sees a huge load, so its size is very big. It is because that the output driver and clock receiver are connected to PAD. The PAD has a large capacitance made by the ESD protection device and package parasitic capacitance. An inductor is also connected between PAD and the package pin. But the output load of replica delay is small. Third, the real path has long metal routing paths whereas the replica delay block has short ones implemented with poly or diffusion resistors due to limited layout area. So, it is important to model the real path precisely. But it is very hard to do precisely because of the mentioned reasons. Therefore, in order to compensate for the mismatch, a post-tuning method is introduced via laser fuse or e-fuse [3–5]. The mismatch between the replica delay and the input and output buffer delays is expressed as gamma as shown in the bottom equation in Fig. 3.5a.

3.2.2 tDQSCK Skew and Locking Consideration

Figure 3.6 shows gamma variation according to supply voltage level in three different process corners in 130-nm CMOS technology. The nominal supply voltage is 1.5V and based on process and voltage variations, the gamma ranges from -95 ps to $+20$ ps. The middle graph in Fig. 3.6b shows the valid data window when gamma is zero. The data valid window is maintained even when subjected to supply voltage

3.2 Basic Concept of DLL

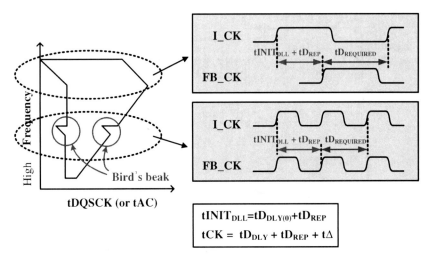

Fig. 3.7 Locking range considerations

variations because the DLL compensates the delay variations. For different values of gamma, the valid window is shifted to the left or right. Designers can correct for the window shift by substituting the gate delay with an RC delay, or vice versa. tDQSCK is adjusted by post tuning method. The left graph in Fig. 3.6b shows the valid data window for tCK variation. As the tCK increases the valid data window increases. For long tCK, the window starts to shift to the left when a lock failure occurs, which is caused by the lack of delay line length.

The lock range of DLL is limited by the total delay length of the variable delay line as shown in Fig. 3.7. For high-frequency operation, the required delay line length is short. In contrast, for low-frequency operation, the needed delay line length is long. The locking range is function of the delay line and replica delay. The initial condition of the DLL before entering the operation is expressed as in these equations. In the case of $N = 2$, unlocked condition occurs when the maximum delay of delay line is less than the tCK period. In case of $N = 2$, the following equations should be satisfied under unlocked condition:

$$0 < tD_{REQUIRED} < tD_{DLY\_LINE\_MAX} \tag{3.5}$$

$$(\text{when}: tINIT_{DLL} > tD_{DLY\_LINE\_MAX}), \tag{3.6}$$

where $tD_{DLY\_LINE\_MAX}$ is the maximum length of the delay line.

We call this condition "the bird's beak" which is expressed as in the bottom equation in Fig. 3.7: where $t\Delta$ is a timing value that is caused by a lack of delay line length. The bird's beak occurs at the condition of $tINIT_{DLL} = tD_{DLY\_LINE\_MAX}$ which is defined as the bird's beak boundary. The bird's beak boundary is defined as the time that the operating tCK period is the same as the length of $tINIT_{DLL}$. The lock range of the DLL varies according to $tINIT_{DLL}$. In case of $N = 1$, the following equations are satisfied under unlocked condition:

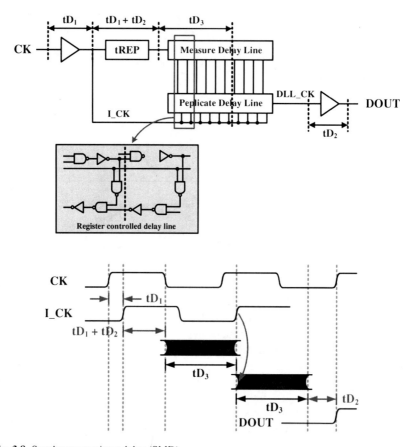

Fig. 3.8 Synchronous mirror delay (SMD)

$$0 < tD_{\text{REQUIRED}} < tD_{\text{DLY\_LINE\_MAX}} + t\text{INIT}_{\text{DLL}} \quad (3.7)$$

$$(\text{when}: t\text{INIT}_{\text{DLL}} \leq tD_{\text{DLY\_LINE\_MAX}}) \quad (3.8)$$

The timing diagram for the lock condition of DLL is shown in the right side of Fig. 3.7. In order to increase the lock range and to solve the initial condition problem, tCK half-period inverting control scheme is used.

3.3 Synchronous Mirror Delay (SMD)

Synchronous mirror delay which has open-loop architecture unlike DLL or PLL is introduced in 1996 [6]. SMD detects the clock cycle from two consecutive pulses to generate a clock synchronized delay as shown in Fig. 3.8. The SMD consists of a replica delay, two delay lines and a control circuit which is represented as black dots of Fig. 3.8. Measure delay line has one input and multiple outputs while replicate

3.3 Synchronous Mirror Delay (SMD)

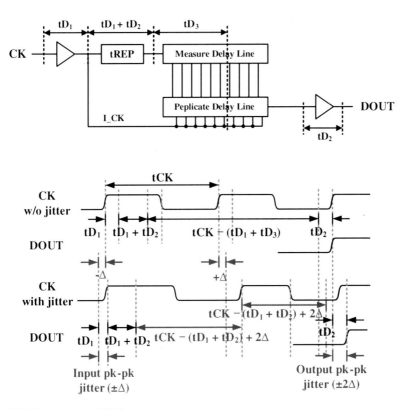

Fig. 3.9 Disadvantage of SMD

delay line has multiple inputs and one output. The two delay lines are connected to each other through the control circuit. Delay cells of the two delay lines are implemented with a NAND gate and an inverter and the black dot is implemented with a NAND gate. The operation of the SMD can be explained using this timing diagram. The input clock goes through the input buffer and replica delay. Then, it goes forward into the upper delay line until I_CK goes into the control circuit. At this time, the clock pulse going through the upper delay line is transferred to the bottom delay line. The bottom delay line replicates the same delay. Then this signal passes through the output driver. Therefore, the phase of signal OUT is synchronized with the input CK after two cycles. In this timing diagram, this part ($tD_1 + tD_1 + tD_2$) and this part (tD_2) vary according to PVT variations. If these parts are increased, then tD_3 will be decreased and vice versa.

Disadvantages of the SMDs are coarse resolution and input jitter multiplication. Let me explain how input jitter is multiplied in detail. As shown in Fig. 3.9, if the input clock has an input jitter of $-\Delta$ at the first cycle, and $+\Delta$ at the second cycle, then the clock period, tCK, is increased by 2Δ. Hence, tD_3 is increased by 2Δ. As a result, the OUT signal has a $+2\Delta$ jitter which means input jitter is multiplied by a factor of two [7].

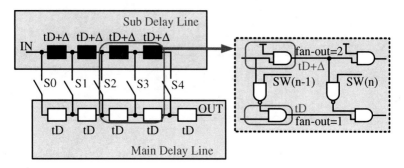

Fig. 3.10 Register controlled DLL

3.4 Register Controlled DLL

3.4.1 Delay Control Method for Register Controlled DLL

Figure 3.10 is a register controlled DLL with venire type delay line. The register controlled DLL includes a shift register to indicate the tap point in the delay line similar to the SMD. The outputs of this register control these switches (SW0~SW4). The SMD from the previous slide has a resolution of tD which is very coarse. To increase the resolution of the delay line, the vernier type delay line consists of two different delay lines. The bottom one has a resolution of tD while the upper one has tD + ∆. Therefore, the resolution is increased to ∆. The left block diagram can be implemented as in the right schematic. The fine delay delta comes from fanout of 2.

The previous SMD and register controlled DLL have two delay lines which are not ideal because of large power consumption. Another method to increase the resolution of DLL using a single delay line is adopting a phase mixer. The resolution of this coarse delay line is a delay through two NAND gates. The phase mixer consists of weighted inverters as shown in the bottom left of Fig. 3.11. Based on the weighting factor, K, the phase of OUT12 can be controlled. If K is 0.5, then OUT12 lies in the middle of OUT1 and OUT2. If K is higher than 0.5, OUT12 lies close to OUT2. By adjusting K, a fine resolution can be achieved. However, this scheme has a boundary switching problem between coarse and fine delay lines.

3.4.2 Boundary Switching Problem

In Fig. 3.12, the DLL is locked within 3 coarse delay cells with K = 0.9. The signal path of this case is shown in solid red line. When the DLL needs to update the locking information to 4 coarse delay cells with K = 0, boundary problem may happen as shown here. Because the coarse shift and fine reset (K reset) do not occur simultaneously, even after the coarse shift left happens, K may still stay at 0.9 for a short time as shown in Fig. 3.12. So we cannot avoid a spontaneous large jitter.

3.4 Register Controlled DLL

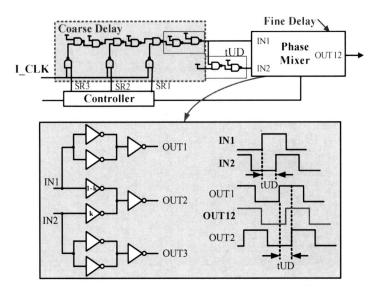

Fig. 3.11 Single register controlled delay line

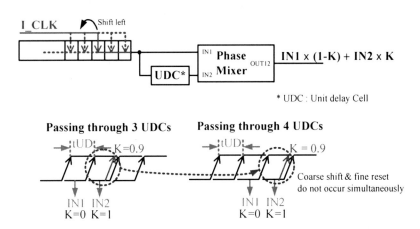

Fig. 3.12 Boundary Switching Problem

To overcome this boundary switching problem, a dual coarse delay line structure was proposed [8]. The upper and lower coarse delay lines consist of odd-numbered and even-numbered coarse delay units, respectively. The outputs of dual delay line feed the phase mixer. In this structure, K is set to 1 first and then the coarse shift happens. K = 1 means that the output of phase mixer is controlled by only IN2. Therefore, the phase shift of IN1 does not contribute any jitter. In the single coarse delay line structure, K value resets abruptly, but in the dual coarse delay line structure, the K value changes continuously between 0 and 1, which enables seamless boundary switching which is shown in Fig. 3.13.

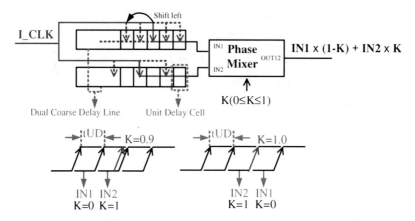

Fig. 3.13 Seamless Boundary Switching

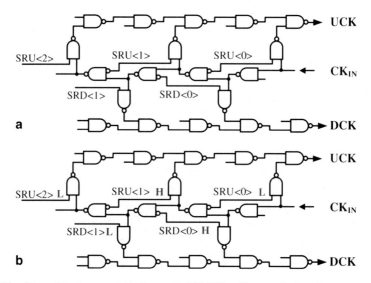

Fig. 3.14 a Merged dual coarse delay line and **b** MDCDL with controller's codes

3.4.3 Merged Dual Coarse Delay Line

Even though the dual delay line solves the boundary switching problem, because of the dual clock path, it requires twice of the power consumption of a single delay line. Furthermore, I_CLK of Fig. 3.11 has many NAND gates to drive. It also dissipates much power. In order to reduce the dynamic power consumption, the merged dual coarse delay line is presented as shown in Fig. 3.14 [9]. The amount of delay to keep the lock condition is determined by the shift register based on the results of the phase detector and the delay controller. The fine mixer interpolates OUT1 and OUT2, whose phase difference is a coarse unit delay length as shown in Fig. 3.14. The

3.4 Register Controlled DLL

weight 'K' controls the interpolation ratio between OUT1 and OUT2. In the merged dual coarse delay line (MDCDL), CK_{IN} is connected to the forward propagation delay path. After that, it splits into two paths: one is a backward propagation path for OUT1 and the other is for OUT2. The signals SRU $< 0:2 >$ and SRD $< 0:1 >$ are the codes for the switch NAND gates. The delay length to be passed is determined by the 'ON' switch NAND gates. If the SRU $< 1 >$ is 'H' and the SRD $< 0 >$ is 'H' then the OUT1 path consists of 6-NAND gates and the OUT2 path consists of 4-NAND gates. These two signals are connected to the fine mixer, and the delay difference between OUT1 and OUT2 is the sum of delay of the fan-out of 2 of one NAND gate and the fan-out of 1 of one NAND gate. In order to increase the delay length, SRD $< 0 >$ goes to 'L' and the SRD $< 1 >$ goes to 'H'. Therefore, the number of NAND gates of the OUT1 path is changed from 4 to 8. The delay difference between OUT1 and OUT2 is also a 2-NAND gate delay. Therefore, four NAND gates are always added to the delay path of each path at every control step, exclusively.

3.4.4 Delay Line Mismatch Issue

The coarse delay is divided into multiple fine delays by a phase mixer. According to K, the fine unit delay is controlled. If the fine step is 4, then the fine unit delay is a quarter of the coarse delay. Even though the coarse delay is designed with the same layout and the same circuits, parasitic mismatch makes the coarse delays differ. Therefore, the fine unit delay is defined as follows:

$$\text{Fine Delay} = \frac{N \times (D+1) - N \times (D+\Delta)}{K \text{ step}} \quad (3.9)$$

where Δ is a mismatch parameter. To simplify the formula, all the upper delay units have the same mismatch parameter. Ideally, the fine delay is $\frac{D}{K\text{Step}}$. If the mismatch of delay unit exists, the fine delay is as follows:

$$\text{Fine Delay} = \frac{D - N \times \Delta}{K \text{ step}} \quad (3.10)$$

If $D < N \times \Delta$, the polarity of the fine delay is negative. Therefore, the delay is not increased according to changing the K step. Finally, the fine step is not a constant value and it is changed by the mismatch parameter of the delay unit.

3.4.5 Adaptive Bandwidth DLL in DRAM

Recently, an adaptive bandwidth DLL with self-dynamic voltage scaling was introduced as shown in Fig. 3.15 [10]. The DLL corrects the delay through the variable delay line when the in/out buffer delays are changed after the locking process. However, it takes time to update the controller output. This update delay is determined by the DLL bandwidth. The bandwidth of the DLL is automatically adjusted according

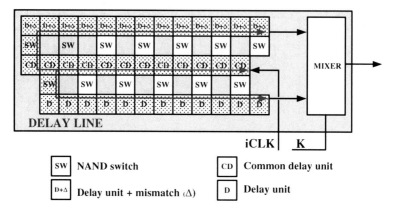

Fig. 3.15 Delay line mismatch effect

to both the process skew condition and the operating frequency by changing the 'm' value. Therefore, the proposed DLL always has the fastest possible tracking speed. The proposed DRAM changes the internal supply voltage to overcome the inevitable process variation problem. It also adjusts the internal supply voltage to reduce power consumption based on the operating frequency needed for the application system. The fine unit delay change based on operating mode is due to self-dynamic voltage scaling. The NCODE is a rounded off binary value of tREP/tCK. tREP is dependent on process variation and is divided by tCK. Therefore, the supply voltage is varied as a function of process variation and tCK. In general, designers over-design their circuit to meet the target requirement even in SS (slow) corner. But by this method, the unnecessary current will be reduced by scaling the supply voltage dynamically.

3.5 CAS Latency Controller

Data is output from DRAM after a predefined number of cycles. These cycles are programmed by an MRS (Mode Register Set) command. This is called Column Address Strobe (CAS) Latency and is defined as the synchronous interval time between the READ command from the MCU and the DQs from the DRAM. In order to issue a READ command with a column address to DRAM, an active command should be issued first with row address. There should be an interval time between a RAS (Row Address Strobe) command and CAS command with the same back. During the initialization sequence, the CL is programmed by an MRS command and the DLL reset command issued. DRAM counts the internal clock cycle to calculate data output timing programmed as CL. One of the output enable control schemes is depicted in Fig. 3.16. The output-enabling time is easily calculated by a simple shift register. However, because the operating frequency goes up and the negative delay is generated by the DLL, it is hard to calculate the output-enabling time through simple shift register. Therefore, output enabling circuits should perceive the time of internal clock delay and negative delay of the DLL. Figure 3.18 shows the countered CAS

3.5 CAS Latency Controller

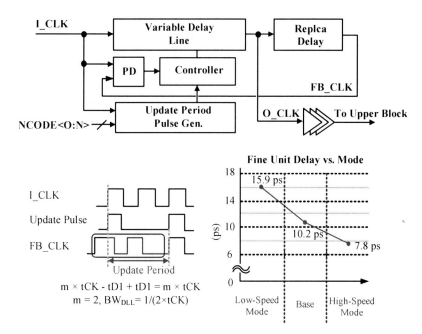

Fig. 3.16 Adaptive bandwidth DLL w/ SDVS

latency control block (CCLC) proposed in another study [9]. The CCLC has an adder, a counter, an OE SHIFT REGISTER, a DELAY MODEL, and an RCDL (register controlled delay line), which is the same as the one in the DLL. The DLL makes a DLL_CLOCK according to the delay model. Eq. (3.9) is defined after locking:

$$tCK \times N = tD_{DLY} + tD_{DELAY\_MODEL}, \quad (3.11)$$

where tD_{DLY} is the delay of the RCDL determined by the delay control block and tD_{DELAY_MODEL} is the delay of the DELAY MODEL. The DELAY MODEL is also the same as the delay model of the DLL.

The real path delay is estimated through DELAY MODEL tD_{DELAY_MODEL} which is an asynchronous value and varies according to the PVT variations. If a DRAM knows the value of tD_{DLY}, CCLC calculates the N of Eq. (3.9). Because the sum of tD_{DLY} and tD_{DELAY_MODEL} is a synchronous value, the value 'N' is easily extracted. Therefore, using a method involving the CCLC subtracting the value 'N' from the CL, CL counting is processed. Finally the number of shift operations in the OE_{SHIFT} REGISTER at a given CL is defined in Eq. (3.10).

$$OE_{SHIFT} = CL - N \quad (3.12)$$

The timing diagram of Fig. 3.19 shows that during the initialization sequence, the CLCC calculates the value 'N' using (3-9). The number in DLL_CLOCK denotes the matched phase with the CLOCK. If the number is 3 in CLOCK, the DLL_CLOCK

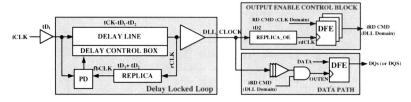

Fig. 3.17 Output enabling block and DLL

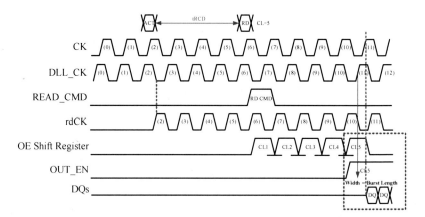

Fig. 3.18 Timing diagram of output enabling

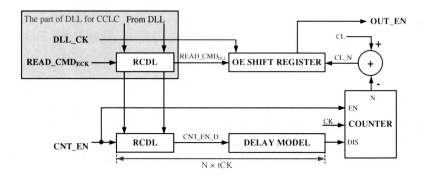

Fig. 3.19 CCLC block diagram

marked 3 is the negative delayed version of CLOCK. If the value N is 3 and CL is set to 5, when the read command is issued at the 6th rising edge of CLOCK, two shift operations occur by the OE SHIFT REGISTER.

The shifting number (OESHIFT) is calculated by 'CL-N' as shown in Eq. (3.10). By adding tRCDL to READ_CMDECK, the rising edges of DLL_CLOCK and READ_CMDDLL are always aligned. Finally, the READ_CMDDLL makes the signal OUT_EN for the burst length period, and the 11th DLL_CLOCK is used to transfer the data.

3.6 Duty Cycle Corrector

Table 3.1 Requirements and design issue of DCC

DCC requirements	DCC design issues
Reduces duty cycle error	Location of DCC (Before/after DCC)
Enlarge valid data window for DDR	Embedded in DLL or not
Needs to correct duty error at maximum speed	Power consumption
	Area
Can be implemented either in analog or digital type	Operating frequency range
	Locking time in case of digital DCC
	Offset of duty cycle detector
	Power down controllability

3.6 Duty Cycle Corrector

DCC is an important building block in DDR memories because DDR SDRAM uses both the rising and falling edges of the clock. As a rule of thumb, in general DCC needs to correct about ± 10 % duty cycle error at maximum speed. It can be implemented either in analog or digital circuitry. When we design a DCC, we need to consider where DCC should be inserted. Some are located before DLL and some are after DLL. Or some are located both before and after DLL. Also some DCCs are embedded in the DLL. Wide operating frequency, locking time and offset of duty cycle detector are also important design parameters. The requirements and design issues of DCC are summarized in Table 3.1.

Digital DCCs can be classified as these three types as depicted in Fig. 3.20. First, the top DCC consists of invert-delay clock generator and phase mixer. The input is inverted and then delayed as shown in the top right timing diagram. Then, the phases of falling edges of the original clock and the inverted clock are mixed to generate a 50 % duty cycle clock. The middle one consists of a pulse width controller and a duty cycle detector. The duty cycle detector detects the duty cycle error and gives feedback to the pulse width controller to adjust the duty ratio. The last DCC consists of a half-cycle delayed clock generator and an edge combiner.

The input is delayed by half cycle and the rising edges of original and delayed clocks are combined through the edge combiner. Top and bottom DCCs are open-loop type and occupies large area due to clock generators and are good for low frequency applications. The middle one is a closed-loop type and needs to control only a small duty error. Therefore, it occupies a smaller area and dissipates smaller power. But it cannot cover a large duty error. Currently, this type is very popular for high-speed memories. One digital DCC is shown in Fig. 3.21 [9]. This DCC is an OR-AND DCC, and it contains OR and AND logics. The OR function widens the high pulse width, and the AND function narrows the high pulse width of the clock. The pulse width can be modulated via CODE $<0:N>$ according to the duty cycle detector. The MSB code of CODE $<0:N>$ is for the direction of duty correction. If the duty of clock needs to be widend, then the OR function path is selected by CODE $<N>$. When the delay of VDL increases by CODE $<0:N-1>$, the overlapped high pulse width between CKD1 and CKD2 also increases. Therefore, the high pulse width of

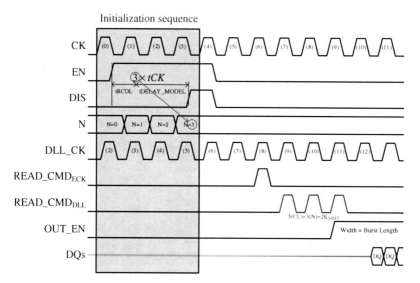

Fig. 3.20 CCLC timing diagram

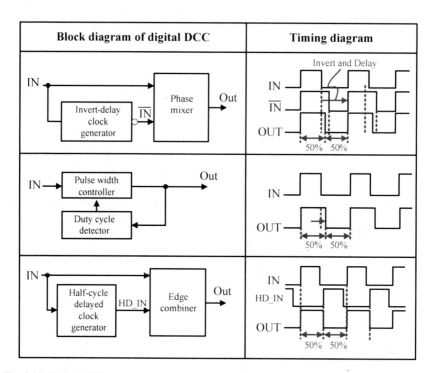

Fig. 3.21 Digital DCC

3.7 DLL Parameters for DRAM

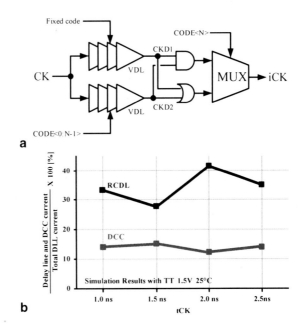

Fig. 3.22 **a** Example of digital DCC and **b** power consumption

iCK increases. In this manner, the AND function narrows the high pulse width of iCK. Because of DCC, the power consumption of DRAM slightly increases. Due to the power-down control, the digital DCC is introduced. The DCC is always operated. Therefore, it is also important to reduce the DCC power. The portion of the DCC block is about 15 % of the total DLL current. This digital DCC consumes less power than the delay line of the DLL.

Figure 3.22 is a DCC used in GDDR5 [11]. As shown in this schematic, this DCC is not on the delay path. So, it does not contribute any additional jitter. In high-speed DRAMS, no jitter addition by DCC is very difficult to achieve. The schematics of this duty cycle adjuster and RX are shown on the bottom right. DCA controls the swing level of either RX output pair. Using 5 bit control, the minimum resolution of this DCC is 6 ps.

3.7 DLL Parameters for DRAM

Table 3.2a lists DLL-related key parameters such as lock time, tDQSCK, speed, tCK and tXPDLL*(tXARD). The lock time of DLL in GDDR is several thousands of cycles while that in DDR is several hundreds of cycles. The operating supply voltage (VDD) for DRAM has been reduced in order to limit power consumption and heat dissipation as the operating frequency is increased. The DLL for DDR3 needs a 512-cycle lock time during its initialization sequence. In order to meet the speed performance requirement, tDQSCK, which is the DQS rising edge output access

Table 3.2 DLL-related parameters

a

	DDR1	DDR2	DDR3	GDDR3	GDDR4
VDD	2.5 V	1.8 V	1.5 V	1.8 V	1.5 V
Lock time	200 cycles	200 cycles	512 cycles	2 ~ 5K cycles	2 ~ 20K cycles
Max. tDQSCK	600 ps	300 ps	225 ps	180 ps	140 ps
Normal speed	166 MHz	333 MHz	333 MHz ~ 800 MHz	600 MHz ~ 1.37 GHz	1.6 GHz
Max. tCK	12 ns	8 ns	3.3 ns	3.3 ns	2.5 ns
tXPDLL (tXARD)	1 × tCK	2 × tCK	Max. (24 ns or 10 × tCK	7 × tCK + tIS	9 × tCK

b

Type	Delay line scheme		DRAM operating state			PSRR	Power consumption (Stand-by + Power down)
	Type	Delay control	IDD2N*	IDD2P*	Self refresh	@Delay line	
Analog	CML	Charge pump	On	On	Off	Good	High
Mixed	CML	Charge pump + ADC	On	Off	Off	Good	Medium
Digital	CMOS	Register controlled	On	Off	Off	Bad	Low

IDD2N*: Stand-by state (Non-power down)
IDD2P**: Power down state (with fast power down exit)

Table 3.3 DLL reference papers

Related area	Reference
DCC block	[13], [14], [16][b], [17], [26][b], [27], [28][b], [29], [31][b]
Variable Delay Line	[14], [18], [20][b], [21], [22][a], [23], [25], [27], [28][b],[29], [30][b], [32], [33]
Delay control logic	[12], [13], [14], [16][b], [17], [18], [19][a], [20][b], [22][a], [23], [24][b], [26][b], [27], [29], [30][b], [31][b], [32], [33][b], [34][a],[35][a]
Replica	[19], [25], [30][b], [33][b]
Low jitter	[14], [15][a], [16][b], [17], [19][a], [21], [23], [25], [26][b],[28][b]

[ ]: Digital type (11 / 24)
[a] Mixed type (5 / 24)
[b] Analog type (8 / 24)

time from the rising edge of CK, becomes an important parameter in the DDR3 system. The tDQSCK variation heavily impacts the performance of the system, especially those with a dual-rank architecture. The GDDR3 speed performance is fast enough to substitute for GDDR4, which is one of the reasons that GDDR4 has hardly been found in the graphics market. The performance of GDDR3 is more than twice that of DDR2. Many of the DLL design technologies created for GDDR3 have been incorporated in DDR3.

The minimum frequency that the DLL needs to meet is increased, which can reduce the delay line length. The parameter tXPDLL is the interval between power down exit and the command requiring a locked DLL. The power-down controllability is determined by the delay line type of the DLL. The CML (current mode logic) type has good PSRR (power supply rejection ratio) characteristics, but the power-down

3.8 Clock Distribution of GDDR5

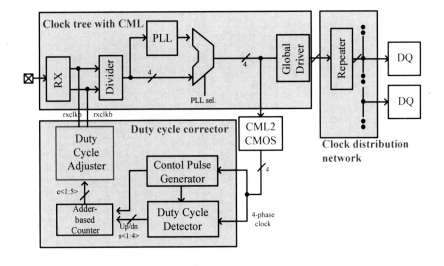

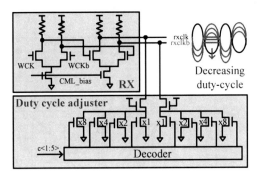

Fig. 3.23 DCC in GDDR5

control is difficult because it is hard to keep the CML bias for adjusting delay. Therefore, an ADC (analog to digital converter) is a good solution. The delay line length which is occupied after locking does not exceed tCK. Higher operating frequency results in a smaller delay being occupied. The digital DLL with a digital delay line is good at high-frequency and with low-power DRAMs.

There have been many papers published to improve the performance of DLL. I have categorized the previous works into five areas according to their topics as shown in the bottom table. Please read those papers for more information.

3.8 Clock Distribution of GDDR5

Figure 3.23 shows the clock configuration for GDDR5. CK and /CK are used for command and address sampling. A command is sampled by SDR, and the address is sampled by DDR. WCK and /WCK are for writing and reading data. The rate

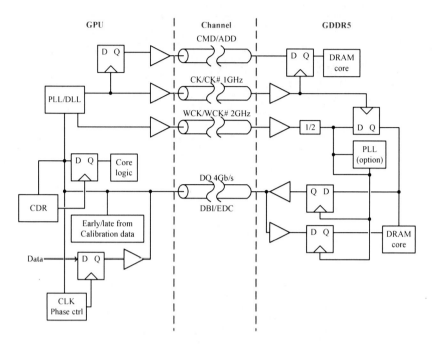

Fig. 3.24 Clock distribution

Fig. 3.25 Clock distribution

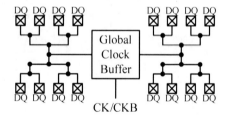

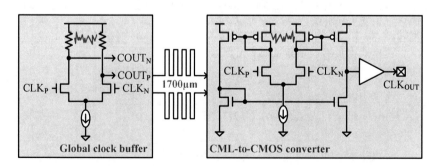

Fig. 3.26 CML to CMOS Converter

of WCK is double CK. The data rate follows the rate of WCK. If the WCK is 4 Gb/s/pin then data is also 4 Gb/s/pin. DBI and EDC stand for data bit inversion and error detection code, respectively. WCK is divided into 4-phase clocks by an internal divider of DRAM. There is no DLL. In the first version of GDDR5, PLL is employed. However, it has been reported that PLL for DRAM is not as good as designers have expected. GDDR5 without PLL is better than one with PLL. DRAM has many power noise sources such as a VPP generator that makes much internal power noise under activation mode, because of low efficiency.

The clock and clock bar are distributed to each DQs by a clock distribution network as shown in the top Fig. 3.24 [36]. To reduce the clock skew among DQs, an H-tree based clock distribution network scheme is common in DRAM. Because the clock distribution network has a huge capacitance and its switching activity is 1, large dynamic power dissipation is inevitable. So, low power global clock buffer is highly demanding. Also, the clock distribution network must be robust under PVT variations, because the clock signal travels through long distances as shown in this die photo. Recently, CML to CMOS converter jitter has becomes a hot issue as well.

Also, the clock distribution network must be robust under PVT variations, because the clock signal travels through long distances as shown in this die photo. Recently, CML to CMOS converter jitter has becomes a hot issue as well.

Recently, the CML buffer is used as the global clock buffer for high-speed memories as shown in Fig. 3.25. This CML clock must be converted to rail to rail swing clock, because the circuits nearby DQs are CMOS circuits. CMOS circuit is more susceptible to supply noise than the CML circuit. For this reason, the CML to CMOS converter circuit generates large jitter. Therefore, it is important to design a noise robust CML to CMOS converter.

References

1. V. Lines et al., "High speed circuit techniques in a 150 MHz 64 M SDRAM," in *Proc. Int'l Workshop on Memory Technology, Design and Testing.*, 1997, pp. 8–11.
2. A. Hatakeyama et al., "A 256-Mb SDRAM using a register-controlled digital DLL," *IEEE J. Solid-State Circuits*, vol. 32, no. 11, pp. 1728–1734, Nov. 1997.
3. S.J. Kim et al., "A low-jitter wide-range skew-calibrated dual-loop DLL using antifuse circuitry for high-speed DRAM," *IEEE J. Solid-State Circuits*, vol. 37, no. 6, pp. 726–734, Jun. 2002.
4. S. Kuge et al., "A 0.18um 256-Mb DDR-SDRAM with low-cost post-mold tuning method for DLL replica," *IEEE J. Solid-State Circuits*, vol. 35, no. 11, pp. 726–734, Nov. 2000.
5. T. Hamamoto et al., "A skew and jitter suppressed DLL architecture for high frequency DDR SDRAMs," in *IEEE Symp. on Very Large Scale Integr. Circuits Dig. Tech. Papers*, 2000, pp. 76–77.
6. T. Saeki et al., "A 2.5 ns clock access 250 MHz 256 Mb SDRAM with a synchronous mirror delay," in *ISSCC Dig. Tech. Papers*, pp. 374–375, Feb. 1996.
7. A. Hatakeyama et al., "A 256 Mb SDRAM using a register-controlled digital DLL," in *ISSCC Dig. Tech. Papers*, pp. 72–73, Feb. 1997.
8. J.-T. Kwak et al., "A low cost high performance register-controlled digital DLL for 1Gbps x32 DDR SDRAM", in *VLSI Dig. Tech. Papers*, pp. 283–284, 2003.

9. H.-W. Lee et al., "A 1.0-ns/1.0-V delay-locked loop with racing mode and countered CAS latency controller for DRAM interfaces," *IEEE J. Solid-State Circuits*, vol. 47, no. 6, pp. 1436–1447, Jun. 2012.
10. H.-W. Lee et al., "A 7.7 mW/1.0 ns/1.35V delay locked loop with racing mode and OA-DCC for DRAM interface," in *Proc. IEEE Int. Conf. Circuits Syst.*, pp. 502-504, May 2010.
11. D. Shin et al., "Wide-range fast-lock duty-cycle corrector with offset-tolerant duty-cycle detection scheme for 54 nm 7 Gb/s GDDR5 DRAM interface," in *VLSI Dig. Tech. Papers*, pp. 138–139, 2009.
12. H.-W. Lee et al., "A 1.6V 1.4 Gb/s/pin consumer DRAM with self-dynamic voltage-scaling technique in 44 nm CMOS technology," *IEEE J. Solid-State Circuits. IEEE J. Solid-State Circuits*, vol. 47, no. 1, pp. 131–140, Jan. 2012.
13. W.-J. Yun et al., "A 3.57 Gb/s/pin low jitter all-digital DLL with dual DCC circuit for GDDR3 DRAM in 54-nm CMOS technology," *IEEE Trans. VLSI Sys., vol. 19, no. 9, pp. 1718–1722, Nov. 2011.*
14. H.-W. Lee et al., "A 1.0-ns/1.0-V Delay-Locked Loop with racing mode and countered CAS latency controller for DRAM interfaces," *IEEE J. Solid-State Circuits*, vol. 47, no. 6, pp. 1436–1447, Jun. 2012.
15. H.-W. Lee et al., "A 1.6V 3.3 Gb/s GDDR3 DRAM with dual-mode phase- and delay-locked loop using power-noise management with unregulated power supply in 54 nm CMOS," in *IEEE Int. Solid-State Circuits Conf. Dig. Tech. Papers*, 2009, pp. 140–141.
16. B.-G. Kim, et al., "A DLL with jitter reduction techniques and quadrature phase generation for DRAM interfaces," *IEEE J. Solid-State Circuits*, vol. 44, no. 5, pp. 1522–1530, May 2009.
17. W.-J. Yun et al., "A 0.1-to-1.5 GHz 4.2 mW all-digital DLL with dual duty-cycle correction circuit and update gear circuit for DRAM in 66 nm CMOS Technology," in *IEEE Int. Solid-State Circuits Conf. Dig. Tech. Papers*, 2008, pp. 282–283.
18. F. Lin, et al., "A wide-range mixed-mode DLL for a combination 512 Mb 2.0 Gb/s/pin GDDR3 and 2.5 Gb/s/pin GDDR4 SDRAM," *IEEE J. Solid-State Circuits*, vol. 43, no. 3, pp. 631–641, March 2008.
19. Y. K. Kim et al., "A 1.5V, 1.6 Gb/s/pin, 1 Gb DDR3 SDRAM with an address queuing scheme and bang-bang jitter reduced DLL scheme" in *IEEE Symp. on Very Large Scale Integr. Circuits Dig. Tech. Papers.*, 2007, pp. 182–183.
20. K.-W. Kim et al., "A 1.5-V 3.2 Gb/s/pin Graphic DDR4 SDRAM with dual-clock system, four-phase input strobing, and low-jitter fully analog DLL," *IEEE J. Solid-State Circuits*, vol. 42, no. 11, pp. 2369–2377, Nov. 2007.
21. D.-U. Lee et al., "A 2.5 Gb/s/pin 256 Mb GDDR3 SDRAM with series pipelined CAS latency control and dual-loop digital DLL," in *IEEE Int. Solid-State Circuits Conf. Dig. Tech. Papers*, 2006, pp. 547–548.
22. S.-J. Bae et al., "A 3 Gb/s 8b single-ended transceiver for 4-drop DRAM interface with digital calibration of equalization skew and offset coefficients," in *IEEE Int. Solid-State Circuits Conf. Dig. Tech. Papers*, 2005, pp. 520–521.
23. Y.-J. Jeon et al., "A 66-333-MHz 12-mW register-controlled DLL with a single delay line and adaptive-duty-cycle clock dividers for production DDR SDRAMs," *IEEE J. Solid-State Circuits*, vol. 39, no. 11, pp. 2087–2092, Nov. 2004.
24. K.-H. Kim et al., "A 1.4 Gb/s DLL using 2nd order charge-pump scheme with low phase/duty error for high-speed DRAM application," in *IEEE Int. Solid-State Circuits Conf. Dig. Tech. Papers*, 2004, pp. 213–214.
25. T. Hamamoto et al., "A 667-Mb/s operating digital DLL architecture for 512-Mb DDR," *IEEE J. Solid-State Circuits*, vol. 39, no. 1, pp. 194–206, Jan. 2004.
26. S. J. Kim, et al., "A low jitter, fast recoverable, fully analog DLL using tracking ADC for high speed and low stand-by power DDR I/O interface" in *IEEE Symp. on Very Large Scale Integr. Circuits Dig. Tech. Papers*, 2003, pp. 285–286.
27. T. Matano et al., "A 1-Gb/s/pin 512-Mb DDRII SDRAM using a digital DLL and a slew-rate-controlled output buffer," *IEEE J. Solid-State Circuits*, vol. 38, no. 5, pp. 762–768, May 2003.

References

28. K.-H. Kim, et al., "Built-in duty cycle corrector using coded phase blending scheme for DDR/DDR2 synchronous DRAM application" in *IEEE Symp. on Very Large Scale Integr. Circuits Dig. Tech. Papers*, 2003, pp. 287–288.
29. J.-T. Kwak, et al., "A low cost high performance register-controlled digital DLL for 1 Gbps x32 DDR SDRAM" in *IEEE Symp. on Very Large Scale Integr. Circuits Dig. Tech. Papers*, 2003, pp. 283–284.
30. S. J. Kim et al., "A low-jitter wide-range skew-calibrated dual-loop DLL using antifuse circuitry for high-speed DRAM," *IEEE J. Solid-State Circuits*, vol. 37, no. 6, pp. 726–734, Jun. 2002.
31. O. Okuda, et al., "A 66-400 MHz, adaptive-lock-mode DLL circuit with duty-cycle error correction [for SDRAMs]" in *IEEE Symp. on Very Large Scale Integr. Circuits Dig. Tech. Papers*, 2001, pp. 37–38.
32. J.-B. Lee et al., "Digitally-controlled DLL and I/O circuits for 500 Mb/s/pin x16 DDR SDRAM," in *IEEE Int. Solid-State Circuits Conf. Dig. Tech. Papers*, 2001, pp. 68–69.
33. S. Kuge et al., "A 0.18um 256-Mb DDR-SDRAM with low-cost post-mold tuning method for DLL replica," *IEEE J. Solid-State Circuits*, vol. 35, no. 11, pp. 726–734, Nov. 2000.
34. J.-H. Lee, et al., "A 330 MHz low-jitter and fast-locking direct skew compensation DLL," in *IEEE Int. Solid-State Circuits Conf. Dig. Tech. Papers*, 2000, pp. 352–353.
35. J. J. Kim, et al., "A low-jitter mixed-mode DLL for high-speed DRAM applications," *IEEE J. Solid-State Circuits*, vol. 35, no. 10, pp. 1430–1436, Oct. 2000.
36. S. J. Kim et al., "A low jitter, fast recoverable, fully analog DLL using tracking ADC for high speed and low stand-by power DDR I/O interface" in *IEEE Symp. on Very Large Scale Integr. Circuits Dig. Tech. Papers*, pp. 285–286, 2003.

Chapter 4
Transceiver Design

In high-speed transceiver design, most of the issues are caused by the lossy channel which has a low-pass filter characteristic. As shown in Fig. 4.1, to compensate for the channel loss, pre-emphasis at the transmitter is performed and equalization at the receiver is performed. Also, the multi-channel interface causes crosstalk and skew effect. These effects should be covered in DRAM interface design. Other issues in transceiver design are impedance matching, low power design, and error reduction. And, as many techniques are applied in the transceiver, training sequence is needed to set the proper operating point of each technique. Several issues and their solutions will be explained in this chapter.

4.1 Lossy Channel

These two types of interfaces have different channels. DDR interface is composed of PCB vias and DIMM socket. Additionally, the connection between CPU and memory data I/O is multidrop. Therefore, the performance of DDR is worse than GDDR. However, GDDR has a point-to-point connection by only PCB lines and vias because GDDR is a performance oriented memory. As the data rate goes higher, the channel between the MCU or GPU and the memory severely attenuates the data signal. The insertion loss of the 8-inch PCB trace is shown in Fig. 4.2. Because of the R and C components in the channel, the channel has low-pass filter characteristics. The insertion losses are −10 dB at 4 GHz, and −15.9 dB at 10 GHz. Depending on the PCB channel length and the trace material, the insertion loss is changed.

As shown in Fig. 4.3, when the original high-speed signal (pulse signal) is transmitted through the lossy channel, transmitted signal may be distorted due to channel loss which causes inter-symbol-interference (ISI). Due to the channel characteristics, the high-frequency components are suppressed, and then, ISI occurs. Under this condition, if the Crosstalk-Induced-Jitter (CIJ) is added while passing the channel, the eye opening of the transmitted data becomes worse as shown in Fig. 4.4. The effect of CIJ is decided by the mutual inductance and capacitance between adjacent channels. When designing a DRAM interface, these effects should be carefully considered.

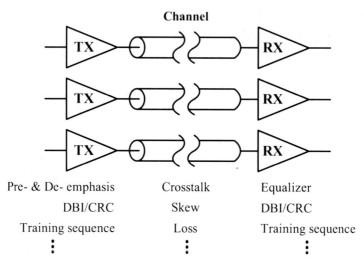

Fig. 4.1 Design issues in the transceiver design

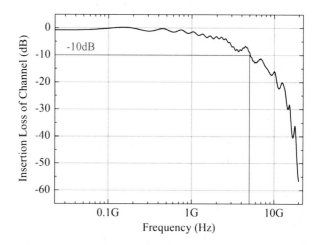

Fig. 4.2 Channel characteristic of the 8-inch PCB trace

4.2 Equalization Techniques

As the data rate has increased to over several Gb/s/pin, channel compensation techniques have become necessary. One of these techniques is emphasis. When the original high-speed signal (pulse signal) is transmitted by the transmitter, this signal may be distorted due to channel loss which causes ISI. Because the channel has a low pass filter characteristic, the high-frequency components are suppressed. As shown in Fig. 4.5, by using the emphasis technique, high-frequency components can be boosted compared to low frequency components. The emphasis technique can compensate for the channel loss. So, we can get a flat frequency response within the target data rate.

4.2 Equalization Techniques

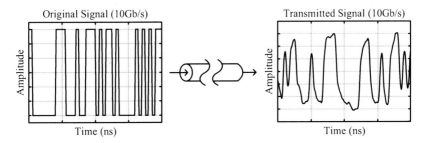

Fig. 4.3 Signal distortion caused by the lossy channel

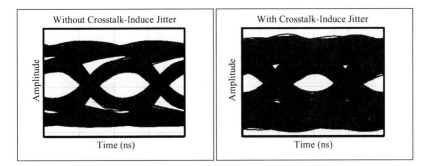

Fig. 4.4 Effect of the Crosstalk-Induced Jitter

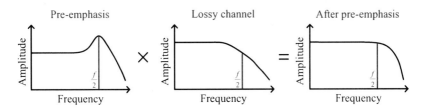

Fig. 4.5 Conceptual operation of the emphasis in frequency-domain

The emphasis technique can be classified as two types as shown in Fig. 4.6. For no emphasis, voltage swing level is Va, The pre-emphasis technique boosts the transition bit which means high-frequency components are boosted while low-frequency components are remained at original voltage level. On the other hand, the de-emphasis technique suppresses the non-transition bit which means low-frequency components are suppressed while high-frequency components maintain the original voltage level. As a result, high-frequency components have relatively more power than low frequency components. Both techniques emphasize high-frequency components compared to low-frequency counterparts and in many case the term of pre-emphasis represents both pre-emphasis and de-emphasis.

Figure 4.7 show common 1-tap de-emphasis techniques. Figure 4.7a shows a mathematical model of the 1-tap de-emphasis. The input data, X(n), and the delayed data is multiplied by K_0 and K_1, respectively, and the multiplied data are summed.

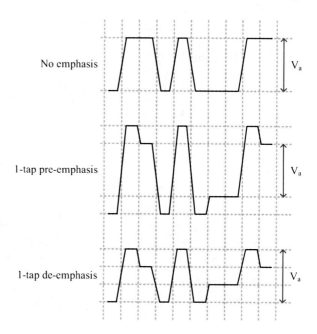

Fig. 4.6 Timing diagram of 1-tap pre-emphasis and de-emphasis

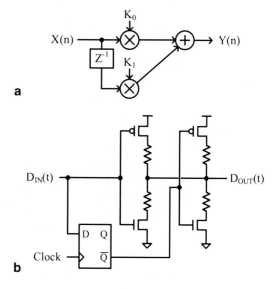

Fig. 4.7 a The mathematical model of 1-tap de-emphasis, and b the schematic of 1-tap de-emphasis

The summer can be designed as a voltage summer or current summer. In the case of a DRAM interface, the voltage-mode output driver with a voltage summer is usually used to reduce power consumption. The overall schematic of the output driver can be realized as shown in Fig. 4.7b. To make one unit delay, a D flip-flop is used and the delayed signal is subtracted from the original signal. In case of the high-speed interface, the variable delay line can be used to make 1UI delay. K_0 and K_1 represent the driver strength and the de-emphasis level can be varied by controlling

4.2 Equalization Techniques

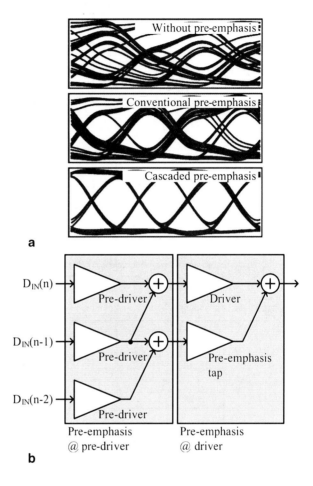

Fig. 4.8 **a** The timing diagram of output data with conventional pre-emphasis and cascaded pre-emphasis, and **b** the block diagram of cascaded pre-emphasis

K_0 and K_1. The number of taps and tap coefficients are changed depending on the channel quality and data bandwidth. To balance efficiency and complexity, one to three taps are generally used.

As shown in Fig. 4.8a, depending on the situation, it may not be enough to apply the pre-emphasis technique at only the final DQ node because internal node would suffer from ISI due to inadequate transistor performance relative to the high data rate. At the external node, the pre-emphasis level and the number of taps should be considered according to the channel quality and length. However, since the load of pre-driver is fixed, the pre-emphasis level and the number of taps can be adjusted with an on-chip calibration. Therefore, if the pre-emphasis which can boost the high-frequency components is applied at the pre-driver, the power consumption of the output driver can be reduced. Based on this fact, the transmitter in a previous study [1], a cascaded pre-emphasis was proposed, which used a pre-emphasis technique at the pre-driver, and its block diagram is shown in Fig. 4.8b. By applying the pre-emphasis technique at the pre-drivers, the power consumption of pre-drivers can be reduced. However, it requires 2UI delayed data for internal pre-emphasis, and the design of the serializer

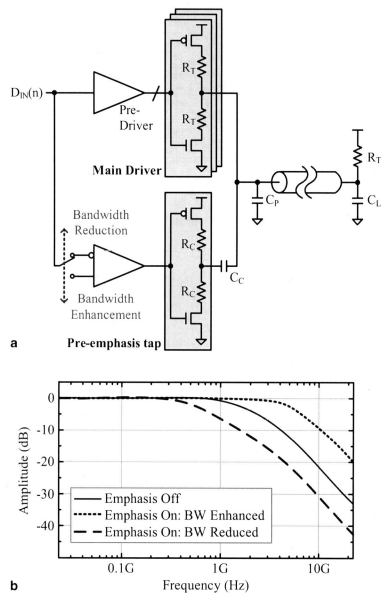

Fig. 4.9 a The schematic of pre-emphasis with zero controlling, and **b** the amplitude of overall transfer function

becomes complex and dissipates additional power. Therefore, the adaptation of the cascaded pre-emphasis should be carefully considered.

Channel loss and compensation technique can be mathematically explained by using poles and zeros. The transfer function of the channel can be simplified with

4.2 Equalization Techniques

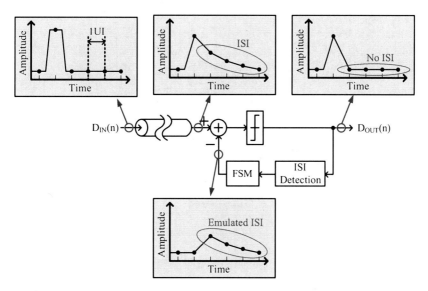

Fig. 4.10 Conceptual block diagram of DFE

multiple poles. Therefore, the channel compensation can be done by inserting a high-frequency zero [2]. Time-discrete pre-emphasis is commonly used in current-mode drivers to minimize the distortion of the data eye due to band limitation of the channel and indicates the presence of a band-enhancing zero. This technique, however, cannot be directly applied to a voltage-mode driver because of the impedance matching problem. As shown in Fig. 4.9a, by using C_C and R_C, an additional zero is controlled, and controlling pre-emphasis tap can enhance or reduce the overall bandwidth of the output driver according to the operating frequency. When the bandwidth needs to be enhanced, the pre-emphasis tap increases the voltage swing. Meanwhile, if the data rate is low, pre-emphasis tap decreases the output voltage swing by subtracting the voltage swing. In this design, the delayed data is not needed. Therefore, the complexity of the serializer can be reduced. The effect of adjusting the zero is shown in Fig. 4.9b. Furthermore, the output of the impedance of the output driver is not changed because C_C acts as an AC-coupling capacitor.

Another equalization technique for memory interface is the decision feedback equalization (DFE) at RX [4]. DFE is more suitable for memory than FFE because memory is single-ended and receives both clock and data, and FFE may require more area and power as well as amplifying high-frequency noise. Figure 4.10 shows a conceptual block diagram of DFE. The channel degrades the input signal depending on its channel loss. And this channel loss creates a long tale called inter-symbol-interference (ISI). To cancel this ISI, DFE generates an emulated ISI depending on previous input data decision. Then, the integrator subtracts the ISI of the received data and slicer decides the input data level. The disadvantage of DFE is that the power of the original data is recovered. Unlike FFE, the power boosting amplifier is not included in DFE. Therefore, to operate the following CMOS circuits, a CML-to-CMOS converter is needed to amplify the signal swing.

Fig. 4.11 a The 1-tap DFE block diagram with a fast feedback time, and **b** the schematic of the DFF sense amplifier

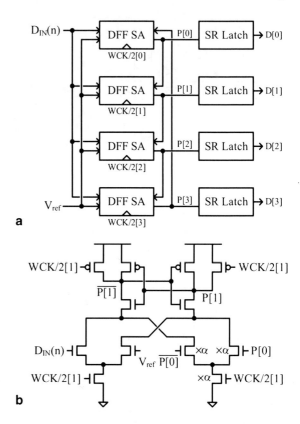

The bottleneck of designing DFE is the timing limitation of the critical path. To subtract ISI components, the decided data should be transferred to the integrator (or summer) before the next data arrives. Therefore, the delay of the feedback path should be smaller than 1UI. As the data rate becomes high, the required timing margin is reduced, but, the delay of the critical path is not reduced. To solve this problem, several techniques are introduced.

Figure 4.11a shows the 1-tap DFE with a fast feedback time in GDDR5 application [4]. The frequency of the sampling clock is WCK/2, and the multi-phase WCK/2 is used. Therefore, the 1UI of data can be defined as one-fourth of WCK/2. As this architecture is a 1-tap DFE, the only first post curser is cancelled. The cancelation is done at the DFF sense amplifier (SA), and its schematic is shown in Fig. 4.11b. In SA, the subtraction coefficient is defined as α. The timing diagram of the fast feedback 1-tap DFE is shown in Fig. 4.12. To reduce the feedback delay, the output of SA is fed back instead of feeding back to an SR latch output. Furthermore, to reduce the delay caused by the aggregating equalizing current, an equalizing path is directly coupled to the SA instead of using a preamplifier. Due to this kind of fast DFE architecture with minimal overhead, this work obtains an operating speed of up to 6 Gb/s/pin. In addition, to enhance the DFE performance, the number of

4.2 Equalization Techniques

Fig. 4.12 Timing diagram of fast feedback 1-tap DFE

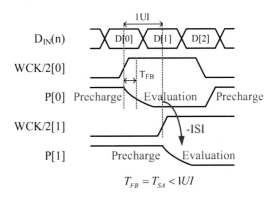

DFE current tap should be increased. However, the memory interface usually has several tens of DQ channels. So if we increase the number of DFE current tap, then the current increases dramatically. Therefore, it is very challenging to increase the number of DFE current tap.

Another DFE is introduced as shown in Fig. 4.13 [5]. In Fig. 4.13a, the delay of the critical path can be reduced by combining a sampler and analog summer. Basically,

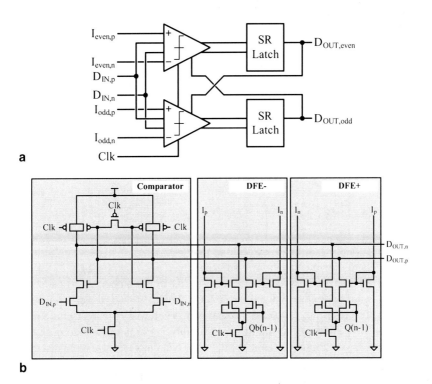

Fig. 4.13 **a** Architecture of DFE, and **b** schematic of comparator

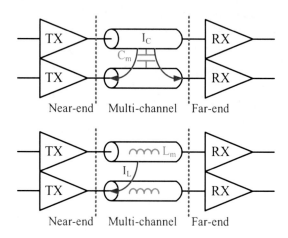

Fig. 4.14 Capacitive and inductive crosstalk

Table 4.1 Effect of crosstalk

Timing	Effect
↗ ↗	Timing Jitter
↗ ↘	Signal Integrity

this architecture has a loop-unrolled predictive DFE without an analog summer. The comparator with DFE taps is shown in Fig. 4.13b. The differential currents, I_p and I_n, are changed due to the comparator offset and the DFE sampling levels. With Fig. 4.13b, the overall power consumption and occupied area can be reduced.

4.3 Crosstalk and Skew Compensation

4.3.1 Crosstalk

Crosstalk is coupling of energy from one line to another. It occurs due to coupled capacitance and inductance between adjacent channels. Figure 4.14 shows such capacitive and inductive crosstalk. As you can see, the crosstalk current by capacitance, I_C, flows toward both near-end and far-end. On the other hand, the crosstalk current by inductance, I_L, flows toward only near-end. Therefore, the near-end current is the sum of these components and the far-end current is the difference of these two components. Crosstalk increases as the data rate becomes faster and the distance between channels becomes narrower. Since multi-channel is used in DRAM to increase the throughput, crosstalk is a leading noise source in high-speed memory interfaces. Table 4.1 shows the effect of the crosstalk. When transitions of two signals occur at the same time, crosstalk between them turns up as timing jitter. If transition is not coincided, crosstalk affects the signal integrity.

4.3 Crosstalk and Skew Compensation

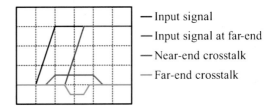

Fig. 4.15 Effect of near-end crosstalk and far-end crosstalk

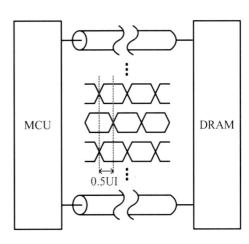

Fig. 4.16 Architecture of a staggered memory bus interface

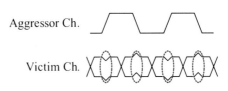

Fig. 4.17 Problem of a staggered memory interface

Figure 4.15 shows the effect of near-end crosstalk (NEXT) and far-end crosstalk (FEXT). When the input signals are transferred through adjacent channels, NEXT occurs with the start of the input signal transition. As the inductive current is dominant at the far-end, FEXT occurs during the transferred signal transition.

One of the methods which can reduce crosstalk is a staggered memory bus as shown in Fig. 4.16 [6]. It is optimized for the case where transitions of adjacent channels occur at the same time. As mentioned above, simultaneous transitions affect timing jitter. In [6], the transition point difference between odd and even channels is a half of the data width. With this, timing jitter caused by same transition timing can be reduced and the distance between channels with the same transition is increased. Therefore, timing jitter by coupling between adjacent channels is reduced.

Unfortunately, a staggered memory has a disadvantage. In actual application, glitch is occurred shown in Fig. 4.17. Glitch is generated due to different transition time and it can affect signal integrity. As shown in Fig. 4.17, rising edge of the aggressor causes the victim signal to decrease instantaneously. On the other hand, the

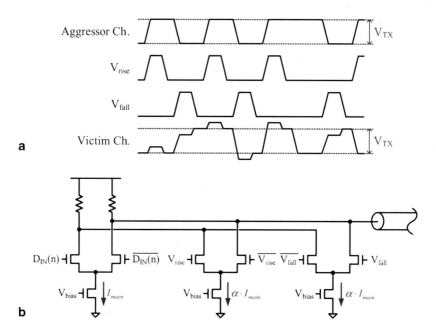

Fig. 4.18 **a** Staggered memory bus with glitch canceller, and **b** the schematic of output driver with glitch canceller tap

falling edge of the aggressor causes the victim signal to increase instantaneously. To remove these glitches, a glitch canceller is applied as shown in Fig. 4.18a. Transition detector adds current to the main driver when the adjacent signal rises. Conversely, it subtracts current from the main driver when the adjacent signal falls. Therefore, the glitch canceller can compensate the glitch. The schematic of the output driver with glitch cancellation is shown in Fig. 4.18b. As the DRAM interface is a single-ended architecture, one output is fed to the channel. If a rising edge or a falling edge is detected, and the channel driving current is summed or subtracted by additional taps.

The other method of crosstalk cancellation is a crosstalk equalizer as shown in Fig. 4.19 [7]. This crosstalk equalization is performed at the transmitter, especially at the output driver. Crosstalk equalizer has several taps to calibrate the output of the main driver. In the control taps, signals of adjacent channels are fed as inputs. Enable signals control this calibration part, and the equalizing coefficient is controlled by enable signals. Then, it cancels the crosstalk by changing the output impedance of the driver.

4.3.2 Skew

Skew is a difference in flight time between signals due to channel disparity as described in Fig. 4.20. Signals which leave from the MCU/GPU at the same time do not arrive at the DRAM simultaneously. This can cause timing errors and this timing error is may cause systems to not operate properly. Also, skew accounts for a greater

4.3 Crosstalk and Skew Compensation

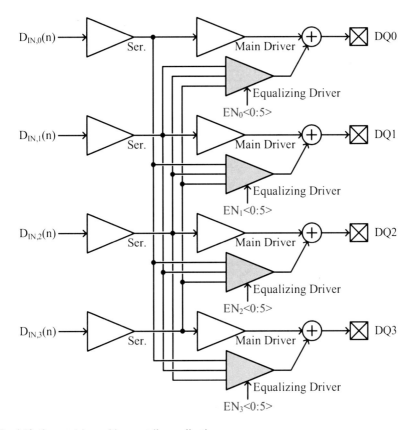

Fig. 4.19 Output driver with crosstalk equalization

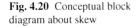

Fig. 4.20 Conceptual block diagram about skew

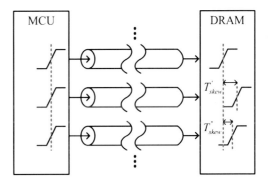

portion in the narrower data unit interval. Therefore, it becomes a more serious problem in high-speed interfaces. So, skew cancellation technique is a key design criterion in high-speed systems.

One method of skew cancellation is pre/de-skew with preamble signal [8]. In this method, skew cancellation circuit is put in each DRAM. Figure 4.21 is a skew cancellation circuit. Before the memory starts to operate, a preamble signal is transmitted

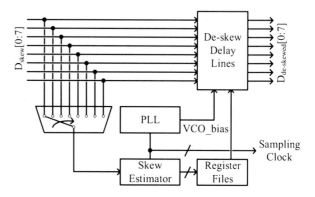

Fig. 4.21 Block diagram of skew cancellation

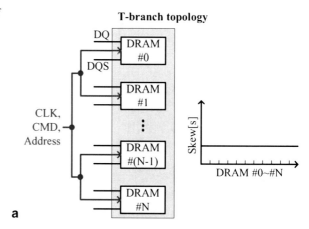

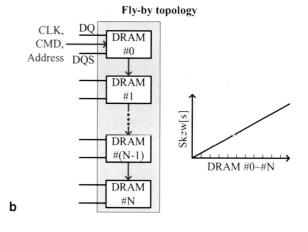

Fig. 4.22 a Block diagram of T-branch topology, and **b** the block diagram of Fly-by topology

from the controller to DRAM. Skew is estimated using this signal and its information is stored in register files. Then, during operation, controller de-skews the data during write mode and DRAM pre-skews the data during read mode. However, this technique consumes large power due to the PLL.

4.3 Crosstalk and Skew Compensation

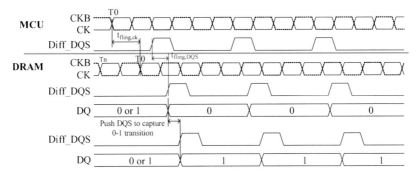

Fig. 4.23 Timing diagram about write leveling technique in DDR3

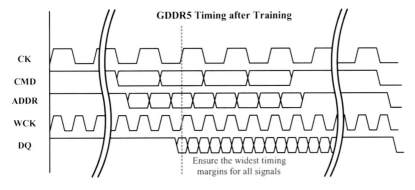

Fig. 4.24 Timing diagram of training sequence in GDDR5

DDR2 uses T-branch topology which supports a parallel connection with each DRAM as shown in Fig. 4.22a. In DDR3, fly-by topology is introduced to reduce the number of stubs and the stub lengths to get better signal integrity at high-speed as shown in Fig. 4.22b. The fly-by topology supports a sequential connection with each DRAM. But this fly-by topology induces large skews according to each DRAM and thus timing problems. To overcome this timing problem, write leveling is introduced which is a kind of training.

To overcome skew problems, write leveling is introduced in DDR3 as shown in Fig. 4.23. When a source (MCU) sends CK and DQS, due to the difference of distance from source (MCU) to each destination, timing problems occur. To solve this, DQS is pushed down to align with CK. When DQS is aligned to CK, DQ driver sends a '1' and by aligning DQS with CK, DQ is also aligned to the clock. This is applied to all DRAMs separately and this timing information is stored in the source (MCU).

GDDR5 supports over 5Gbps data rate communications. At this high data rate, a slight timing mismatch can cause communication errors. To reduce this timing problem at very fast data rate, the training method is introduced for GDDR5. These trainings consist of address, clock, read and write training. As shown in Fig. 4.24, the trainings are for ensuring the widest timing margins on all signals. These trainings are controlled by the MCU and after training the results are saved in the MCU.

Fig. 4.25 Training sequence of GDDR5

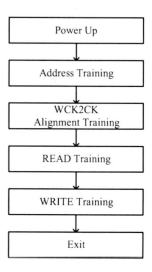

The training sequence is shown in Fig. 4.25. And its explanation is as below.

1. During Power up, the device configuration (x32/x16 mode) and mirror function are detected. In addition, the ODT for the address/ command lines are set.
2. The address training may be used to center the address input data eye.
3. WCK2CK alignment training. WCK and CK clocks require a certain relationship. This phase relationship ensures a reliable phase-over of write data and read data.
4. Read and write trainings are for searching for the best read/write data eye. First, these trainings align the data bus to the WCK clock. And then the burst boundaries out of a continuous read/write data stream will be detected.

Figure 4.26 show the WRITE training example for GDDR5. In this case the GPU may transmit data to each DRAM cell at the same time. The 1UI width is very narrow, so a small driver and channel mismatch can cause skew between each DQ channels as shown in Fig. 4.26a. In this case, the GDDR5 device cell cannot sample the transmitted data properly. The skew can be compensated by the WRITE training as shown in Fig. 4.26b. Now, if write training is supported, these timing mismatches are inversely added at the GPU side. As a result, at the GDDR5 device side, the arrival time of each DQs can get the same data timing.

4.4 Input Buffer

Input buffer is the first stage that receives and senses the external signal as shown in Fig. 4.27. The external signals include clock, data, command and address. The pin composition can be changed according to the DRAM type. The input buffer converts the attenuated external signal to rail-to-rail signal. Because input buffers consume significant power, and faster operation speed requires greater power consumption,

4.4 Input Buffer

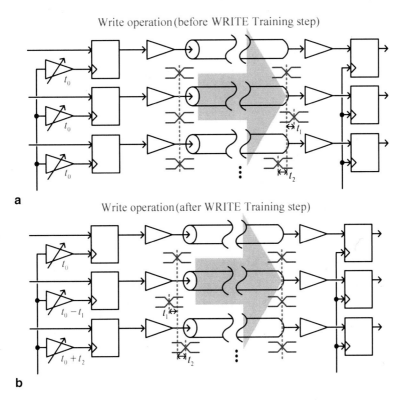

Fig. 4.26 **a** Example for WRITE training step of GDDR5 before WRITE Training step, and **b** that after WRITE Training step

Fig. 4.27 DRAM interface with location of input buffer

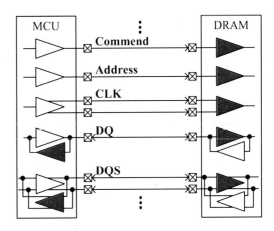

balancing the trade-offs between high speed operation and power consumption is very important.

Two kinds of input buffers are used in DRAM as indicated in Fig. 4.27. In the case of the CLK and DQS pins, the differential-type input buffer is used. In contrast, the single-ended input buffer is used for the DQ, command, address pins.

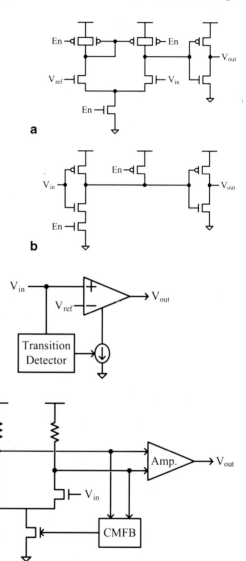

Fig. 4.28 Schematic of **a** differential-type input buffer, and **b** single-ended input buffer

Fig. 4.29 Schematic of **a** input buffer with transition detector, and **b** input buffer with CMFB

Figure 4.28a is differential type and Fig. 4.28b is CMOS Type. Differential type is more popular, and each type has its own pros and cons. Compared to CMOS type, differential type input buffer is complex but better for high-speed operation. Moreover, it is robust to noise and has a stable threshold. The CMOS type input buffer circuit is simple but its speed is lower. Also it is more susceptible to noise and has an unstable threshold.

As the insertion loss increases due to the increment of data rate, the eye opening of the received data is not enough to satisfy voltage and timing margin. Therefore, the gain enhanced buffer and wide common mode rage DQ input buffer were introduced [9]. In Fig. 4.29a, the transition detector provides additional bias current. By

4.5 Impedance Matching

Fig. 4.30 The mathematical model of impedance matching

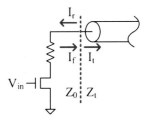

adding current when the transition of input data is detected, the gain enhancement is achieved. Furthermore, as shown in Fig. 4.29b, the wide common mode range DQ input buffer has a feedback network to reduce the output common mode variation. This front-stage delivers stable inputs to the second stage amp.

4.5 Impedance Matching

Impedance mismatching causes reflection, which increases the jitter of the transmitted data. Because of this, impedance matching is an important issue in high-speed signaling. The effect of the impedance mismatching is calculated by Telegrapher's equation as in the equation below and Fig. 4.30:

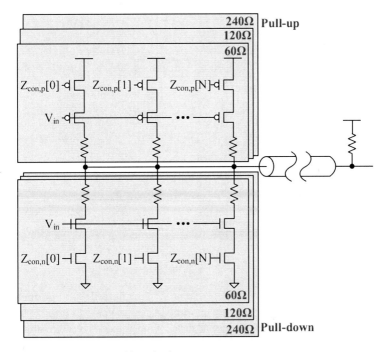

Fig. 4.31 Basic impedance matching circuit

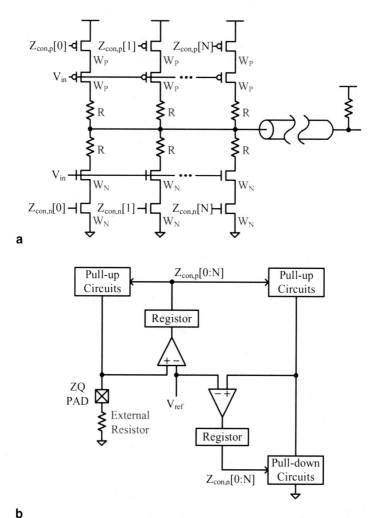

Fig. 4.32 a Schematic of unit block for impedance matching, and **b** the impedance calibration logic

$$I_t = \frac{2V_i}{Z_0 + Z_t}, I_r = I_f - I_t$$

$$I_r = \frac{V_i}{Z_0}\left(\frac{Z_t - Z_0}{Z_t + Z_0}\right)$$

As shown in this equation, when Z_T is same as Z_0, the reflection current disappears. Therefore, in the DRAM interface, impedance matching circuits should be included. The basic single-ended output driver with impedance matching circuits is depicted in Fig. 4.31. The impedance matching circuit consists of several unit impedance blocks which have values of 60ohm, 120ohm, and 240ohm to produce

4.5 Impedance Matching

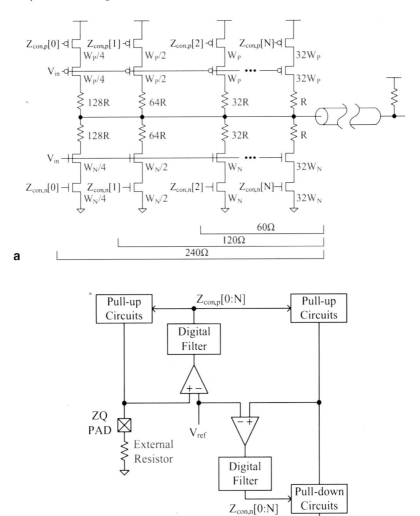

Fig. 4.33 a Schematic of unit block for multi slew-rate controlling, and b its control logic

various impedances. Parallel unit blocks produce 120ohm or 60ohm depending on the required impedance. The impedance calibration of each block is done by controlling $Z_{con,p}[0:N]$ and $Z_{con,n}[0:N]$ signals. It can be manually calibrated or automatically calibrated. Furthermore, the impedance matching of pull-up circuits and pull-down circuits is operated independently, and the matching resistor in RX is usually connected to VDD.

Each unit block in the output driver with impedance matching can be designed as Fig. 4.32a [10]. In this structure, each lag of the unit block which consists of a MOSFET and a resistor has the same size. The impedance of the output driver is controlled with Fig. 4.32b. First, pull-up impedance is controlled based on the external resistor and the reference voltage. And then, pull-down impedance is controlled

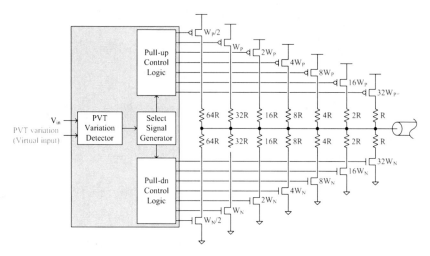

Fig. 4.34 Impedance matching circuit without the external resistor

based on the calibrated pull-up impedance and reference voltage. The lags in the unit block are controlled by thermometer code which is stored in a register. This scheme has a limited calibration range.

To increase the calibration range with a given control bit, binary-weighted control method can be used as shown in Fig. 4.33 [11]. This structure has a large calibration range. So, a unit block with binary-weighted sized lags which is shown in Fig. 4.33a makes it possible to control multi slew-rate with reduced number of control bits compared to thermometer code control method. Overall operation is similar to Fig. 4.32. Compared to the Fig. 30, the difference is a binary-weighted code control. Figure 4.32a uses a register to control the impedance. In contrast, Fig. 4.33b uses a digital filter composed of digital low pass filter and up/down counter. With this method, we can obtain matching impedance. However, in the DRAM interface, there are many DQ pins, and the impedance of each DQ pin is controlled by one calibration circuit. Due to the PVT, the impedance of each DQ pin has a different variation. As a result, the impedance of each DQ pin cannot be calibrated properly.

The previous impedance matching circuits need an external resistor. However, slew-rate calibration can be done without an external resistor as shown in Fig. 4.34 [12]. PVT variation is detected by the PVT variation detector, and it makes it possible to obtain constant slew-rate, which means that the output driver has a constant impedance. The advantage of this structure is an open-loop structure and its lock time is only one clock period. The size of MOSs and resistors are similar to those in Fig. 4.32a.

In the memory cell, only one pad, ZQ, is allocated for the external resistor. However, if X32 IOs are used, global ZQ calibration is needed. In [13], proposes global ZQ calibration with additional blocks as shown in Fig. 4.35. In this design, local PVT sensor finds PVT variation, and, using this sensor and local controller, it calibrates impedance of each IO PAD. As a result, global impedance mismatch is less than 1 %.

4.6 DBI and CRC

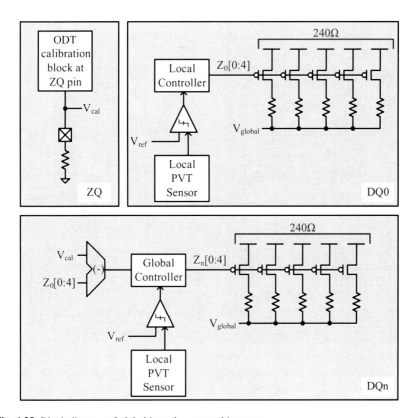

Fig. 4.35 Block diagram of global impedance matching system

Otherwise, the impedance matching can be intentionally broken to reduce power consumption of the TX as shown in Fig. 4.36a [14]. As I/O interface circuits consume around 20~30 % of the overall power, the increment of resistance of the output driver effectively reduces the overall power consumption. However, reflection occurs due to the impedance mismatch between the output driver and channel. In [14], the reflection is cancelled by measuring the amount of reflection and the position of the reflection signal, as shown in Fig. 4.36b. With the help of a reflection cancelling circuit, the overall power consumption can be reduced by 27 %. However, as the position of the reflection signal is decided by the number of flip-flops, the cancelling range is limited.

4.6 DBI and CRC

Data Bus Inversion (DBI) is adopted for power savings that is independent of the data pattern. As shown in Fig. 4.37, when the 8bit data is fed, a 'Low(0)' counter counts the number of '0' value in input data. If it is over 4, DBI signal will change

Fig. 4.36 a Conceptual circuit, and **b** the block diagram which is introduced in [12]

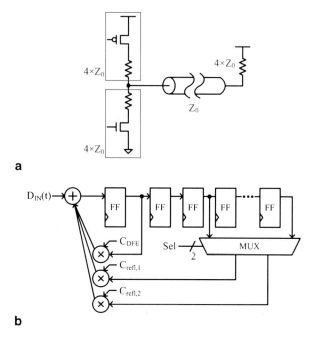

to '1' and the DQ drivers will send inverted DQ data. As a result, the number of 0 data can be reduced to fewer than 4. I/O buffer power dissipation can be reduced significantly compared with no DBI scheme. However this technique requires one more pad to transfer the DBI. Not only that, this inversion time + Cyclic Redundancy Check (CRC) time can be a bottleneck of speed limit.

In practice, 8 CRC bits are added for 64 received data and 8 DBI bits as shown in Fig. 4.38a [15]. To calculate CRC[0], 33 XOR logics are used as expressed as Fig. 4.38b. Because of this complex logic, XOR logic optimization is needed to increase the CRC speed. Data is selected by using the polynomial expression $X^8 + X^2 + X + 1$ with initial value of '0'. In CRC0, 34 data out of 72 (64bit data + 8bit DBI) are selected to detect the error. If any error is detected, the memory module will notify the error to the GPU or CPU and receive the corrupted data again, and recover it.

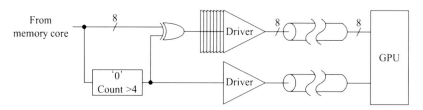

Fig. 4.37 Block diagram of DBI

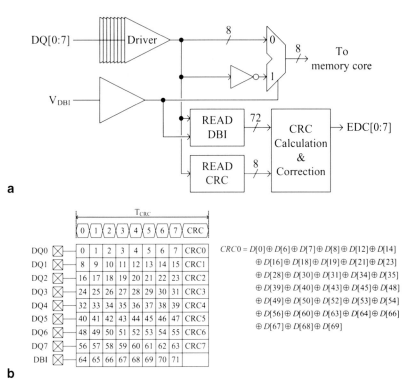

Fig. 4.38 **a** Block diagram of CRC logic, and **b** CRC calculation algorithm

References

1. K.-h. Kim et al., "A 20-Gb/s 256-Mb DRAM with an inductorless quadrature PLL and a cascaded pre-emphasis transmitter," *IEEE J. Solid-State Circuits*, vol.41, no. 1, pp. 127–134, Jan. 2006.
2. H. Partovi et al., "Single-ended transceiver design techniques for 5.33 Gb/s graphics applications," in *IEEE ISSCC Dig. Tech. Papers*, Feb. 2009, pp. 136–137.
3. Y. Hidaka, "Sign-based-Zero-Forcing Adaptive Equalizer Control," in *CMOS Emerging Technologies Workshop*, May 2010.
4. S.-J. Bae et al., "A 60 nm 6 Gb/s/pin GDDR5 graphics DRAM with multifaceted clocking and ISI/SSN-reduction techniques," in *IEEE ISSCC Dig. Tech. Papers*, Feb. 2008, pp. 278–279.
5. K. Kaviani et al., "A 6.4 Gb/s near-ground single-ended transceiver for dual-rank DIMM memory interface systems," in *IEEE ISSCC Dig. Tech. Papers*, Feb. 2013, pp. 306–307.
6. K.-I. Oh et al., "A 5-Gb/s/pin transceiver for DDR memory interface with a crosstalk suppression scheme," *IEEE J. Solid-State Circuits*, vol. 44, no. 8, pp. 2222–2232, Aug. 2009.
7. S.-J. Bae et al., "A 40 nm 2 Gb 7 Gb/s/pin GDDR5 SDRAM with a Programmable DQ Ordering Crosstalk Equalizer and Adjustable clock-Tracing BW," in *IEEE ISSCC Dig. Tech. Papers*, Feb. 2011, pp. 498–500.
8. S. H. Wang et al., "A 500-Mb/s quadruple data rate SDRAM interface using a skew cancellation technique," *IEEE J. Solid-State Circuits*, vol. 36, no. 4, pp. 648–657, Apr. 2001.
9. K. Sohn et al., "A 1.2V 30 nm 3.2 Gb/s/pin 4 Gb DDR4 SDRAM with dual-error detection and PVT-tolerant data-fetch scheme," in *IEEE ISSCC Dig. Tech. Papers*, Feb. 2012, pp. 38–40.

10. C. Park et al., "A 512-mb DDR3 SDRAM prototype with CIO minimization and self-calibration techniques," *IEEE J. Solid-State Circuits*, vol. 41, no. 4, pp. 831–838, Apr. 2006.
11. D. Lee et al., "Multi-slew-rate output driver and optimized impedance-calibration circuit for 66 nm 3.0 Gb/s/pin DRAM interface," in *IEEE ISSCC Dig. Tech. Papers*, Feb. 2008, pp. 280–613.
12. S. Kim et al., "A low-jitter wide-range skew-calibrated dual-loop DLL using antifuse circuitry for high-speed DRAM," *IEEE J. Solid-State Circuits*, vol. 37, no. 6, pp. 726–734, Jun. 2002.
13. J. Koo et al., Young Jung Choi, and Chulwoo Kim, "Small-Area High-Accuracy ODT/OCD by calibration of Global On-Chip for 512M GDDR5 application," in *IEEE CICC*, Sep. 2009, pp. 717–720.
14. S.-M. Lee et al., "A 27 % reduction in transceiver power for single-ended point-to-point DRAM interface with the termination resistance 4xZ0 at TX and RX," in *IEEE ISSCC Dig. Tech. Papers*, Feb. 2013, pp. 308–309.
15. S.-S. Yoon et al., "A fast GDDR5 read CRC calculation circuit with read DBI operation," in *IEEE Asian Solid-State Circuits Conf.*, Mar. 2008, pp. 249–252.

Chapter 5
TSV Interface for DRAM

This topic is Thru Silicon Via interface for DRAM. This part will explain why TSV is necessary in DRAM, the advantages of DRAM with TSV, the TSV DRAM types and the issues and solutions for the TSV DRAM.

5.1 The Need for TSV in DRAM

Figure 5.1 shows the DRAM data rate/pin trends. From 200 Mb/s/pin to 7 Gb/s/pin, the data rate has increased significantly during the last decade. If this trend is to follow, the data rate/pin of the next GDDR will require over 10 Gb/s/pin. However, it is very difficult to achieve a bandwidth over 10 Gb/s/pin with single-ended signaling using DRAM process technology. Even if the data rate reaches 10 Gb/s/pin, the additional circuits such as clock and data recovery circuit (CDR) and continuous-time linear equalizer (CTLE) will increase active area and power consumption [1]. For this reason, DRAMs with TSV are developed. Power consumption of four stacked TSV chips is lesser than that of QDP according to one study as shown in Fig. 5.2 [2]. IDD2N represents the standby current with no power-down mode. All Rx circuits need to be activated while receiving command. Also, the DRAM core circuit (for cell control) is under a pre-charge state. In the case of TSV DRAM, the number of Rx circuits and DLL is less than those of QDP. Therefore, the IDD1 (Active current) is also reduced. Moreover, the operating speed increases from 1,066 Mb/s/pin to 1,600 Mb/s/pin.

5.2 Die Stacking Package

There are three major stacking type packages. The multi-chip package is a popular package to stack different types of dies. Stacking dies are employed to increase the chip I/O density and performance. A general MCP typically consists of 2–6 dies and is packaged by package on package or die on die. Without increasing the density and performance of a single chip, MCP achieves a small form factor. Therefore, it reduces the board area and routing complexity. This is one of the cost effective solutions.

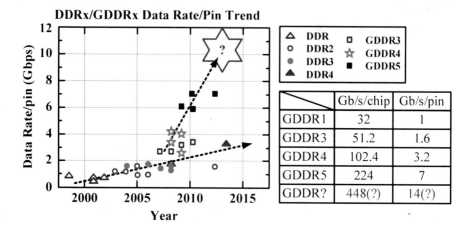

Fig. 5.1 Bandwidth requirements

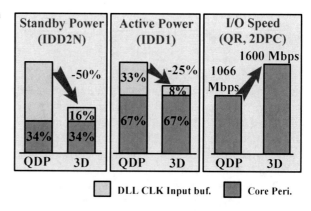

Fig. 5.2 Power consumption comparison between QDP and 3D chip

A multi-chip package has 2 ~ 24 dies encapsulated to increase density and performance. It is processed in two structures: vertically positioned or stacked chips and side-by-side positioned chips. The former needs less surface mount area. They are employed in mobile devices for compact size. The latter is for multi functioned chips.

Figure 5.3 shows the cross section of three kinds of package types for multi-chip encapsulation.

5.3 Configuration of DRAM and MCU/GPU via TSV

As shown in the Fig. 5.4, memories can be stacked and connected through silicon vias. An interposer is used to connect the MCU with memory. A DRAM with TSV has several advantages such as high density, low power and fast flight time. Additionally, the DRAM with TSV can reach higher bandwidth with wider I/O. Therefore, only 800 Mb/s/pin with 512 I/O will achieve 448 Gb/s/chip at next generation GDDR.

5.4 Types of TSV-Based DRAM

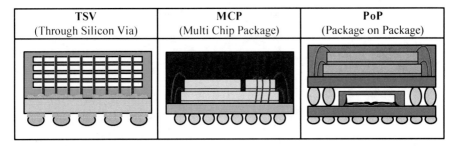

Fig. 5.3 Cross-section according to package type

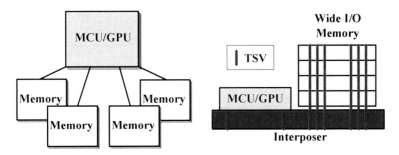

Fig. 5.4 DRAM with TSV

Type	Main Memory	Mobile	Graphics
Architecture	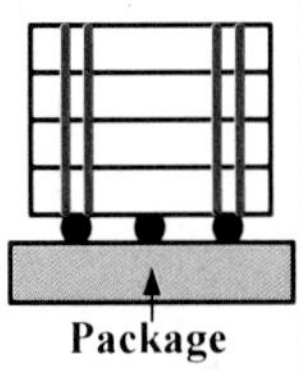 Package	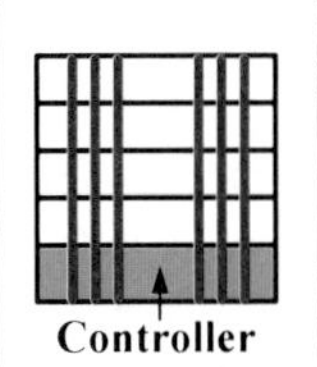 Controller	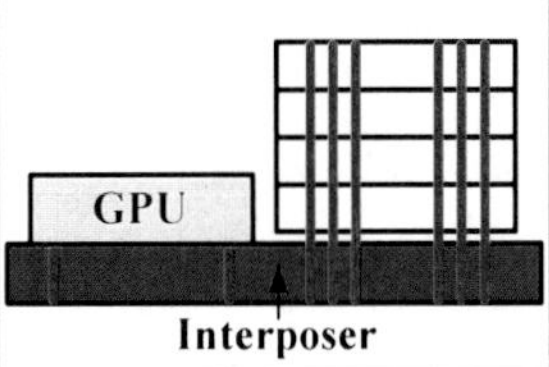 Interposer
No. of TSV	500~1000 EA	1000~1500 EA	2000~3000 EA
Feature	• Low power • High speed	• Low power • Multi channel • Wide I/O	• Max bandwidth • Multi channel

Fig. 5.5 TSV DRAM type

5.4 Types of TSV-Based DRAM

TSV DRAMs have architecture variations according to the applications as shown in Fig. 5.5. In the main memory use case, the stacked memories are connected to the package directly and the number of TSVs is 500 ~ 1,000. In the mobile use case, the memories are stacked with controller chip and the number of TSVs is 1,000 ~ 1,500. In the graphics use case, the stacked memories and GPU are connected by an interposer and the number of TSVs is 2,000 ~ 3,000. Even though the TSV DRAMs

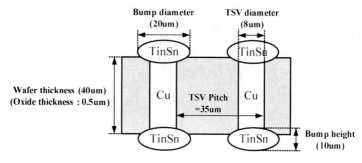

TSV Diameter	Wafer Thickness	TSV Pitch	Oxide Thickness	Bump Diameter	Bump Height
8μm	40μm	35μm	0.5μm	20μm	10μm

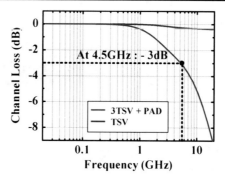

Fig. 5.6 TSV characteristics

have minor differences in architecture and in the number of TSVs, the trend of the overall memory applications is the same: high bandwidth and low power.

One example of TSV structure is presented in Fig. 5.6 [3]. In general, TSV-based DRAM has wide I/O and lower speed performance. The channel loss of TSV is less than 1 dB under a 1 GHz of operating frequency. Due to the wide geometry, the resistance is small and the capacitance is large.

There are two stacking types, homogeneous and heterogeneous. Homogeneous type stacks the same kind of chips. Therefore, the homogeneous type is cheaper than heterogeneous. Heterogeneous type stacks the master and slave chips. In general, the slave chips consist of only memory cells and the master chip consists of cells with peripheral circuit, as shown in Fig. 5.7.

Fig. 5.7 Stacking type

Type	Homogeneous	Heterogeneous
Architecture		Slave / Slave / Slave / Master
Feature	• Same chips • Low cost	• Slave : only cells • Master : with peripheral

5.5 Issue of TSV-Based DRAM

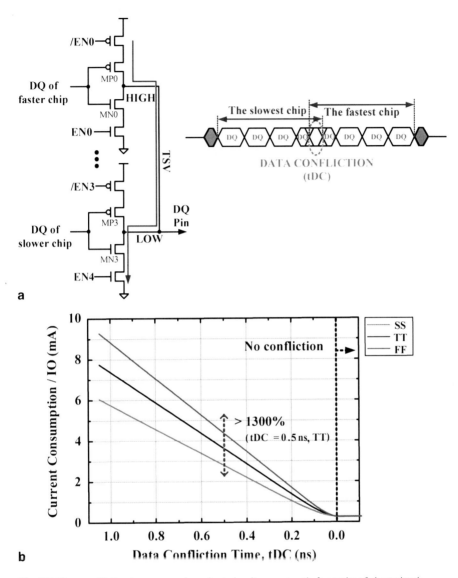

Fig. 5.8 Data confliction issue example **a** short circuit current path, **b** graphs of short circuit

5.5 Issue of TSV-Based DRAM

Nevertheless, DRAM with TSV needs to overcome some issues. One of the issues is data confliction [3] as shown in Fig. 5.8a. The data confliction issue is caused by the PVT variations among stacked chips and shared I/Os. As shown here, if one die

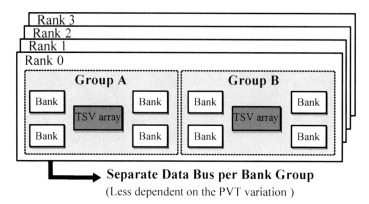

Fig. 5.9 Separate data bus per group

gets slower and the other gets faster, then data conflicts like this. This issue also increases the short circuit current. For example, when CHIP3 is driving a logic low data, if CHIP0 is driving a logic high data, short circuit current will flow from MP0 to MN3. This phenomenon has been overcome by several solutions. Figure 5.8b shows the simulation results of current consumption when data confliction occurs. Over 1,340 % excessive current is confirmed at 0.5 ns of tDC. Because of the larger tDC variances among the TSV-based stacked dies, inevitable power dissipation caused by short circuit path occurs. In general, the number of IOs for TSV DRAM might be 512 or more [3, 4], and it is easy to expect huge additional short circuit power dissipation. Even though mobile DRAM does not need the DLL because of the power budget [3], TSV DRAM for high-speed operation with the DLL in a master die was presented in [5]. Because of the process variation, each die affects the circuit performance significantly. Therefore, the data confliction of TSV-based stacked dies not only makes poor valid data window, but also additional power consumption.

Separated data bus per bank group is one solution [2] as shown in Fig. 5.9. The short distance from bank to the TSV array helps to lessen the impact of PVT variation. For this reason, the output data timing is nearly the same among stacked chips under PVT variation. Therefore, the data confliction can be minimized.

Another solution is to use a DLL-based data self-aligner scheme [3]. This scheme aligns the data of each die to the reference clock. The reference clock can be change to external clock or clock of the slowest chip according to the operation mode. If data is aligned to external clock, the synchronous mirror delay type DLL (skew detector and skew compensator) aligns the clock to external clock with fast lock time. However, in the mode where the data is to be aligned to the clock of the slowest chip, the fine aligner and the PD2 align the clock with low power consumption because this mode does not use the SMD type DLL.

In Fig. 5.10, one of the TSV-stacked dies is selected to process the read operation and others enter the snoop read operation. This is determined by a read command with

5.5 Issue of TSV-Based DRAM

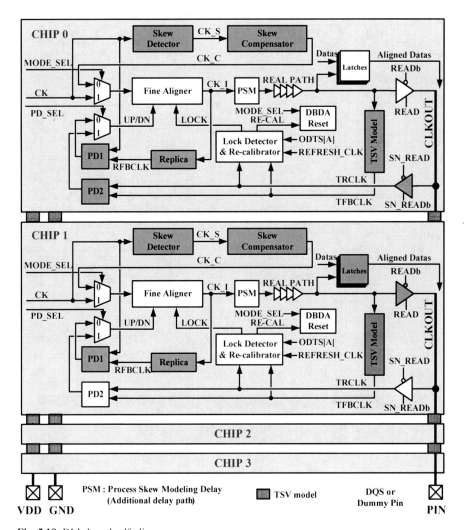

Fig. 5.10 DLL-based self-aligner

non-activated/CS. When a die is under a snoop read mode, DBDAs are awakened and start to align TFBCLK to TRCLK by the fine aligner.

Assume that CHIP0 enters the read operation and CHIP1 ~ 3 enter the snoop read operation. The CLKOUT of CHIP0 is the reference clock for other pseudo-stacked dies. CHIPs 1 ~ 3 start to align each TFBCLK to TRCLK from CLKOUT. The fine aligner can only increase the delay of the delay line, the fine aligners of CHIP1 and CHIP3 are in waiting mode. After that, CHIP1 enters the read operation and CHIP0 and CHIP2 increase the delay of the fine aligner after that. According to these sequences, all TFBCLKs are aligned to the CLKOUT. Finally, all clocks are aligned as shown in Fig. 5.11. CHIP3 is the slowest chip among the TSV-stacked

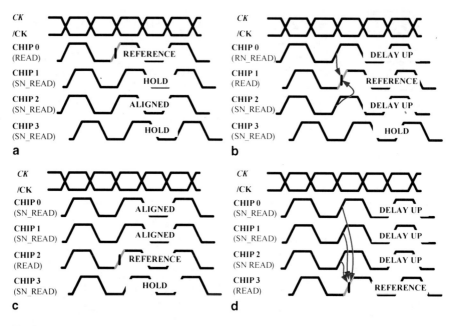

Fig. 5.11 Timing diagram of proposed DBDA operation at **a** CHIP 0 = READ, **b** CHIP 1 = READ, **c** CHIP 2 = READ and **d** CHIP 3 = READ

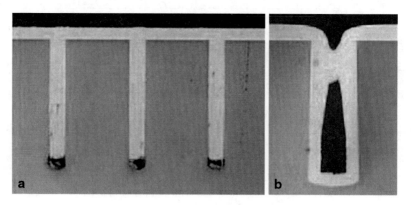

Fig. 5.12 Examples of failed TSV with **a** bottom plating voids **b** a pinch-off

dies. Regardless of the order of the read sequence and the location of the slowest die, the DBDAs always align their tQ to the slowest die's tQ. Therefore, tDC is removed by the DLL-based data self-aligner.

Another issue to consider is a failed TSV. Figure 5.12a illustrates a TSV plating defect, which is bottom plating voids [6]. And Fig. 5.12b shows a pinch-off, which means that via is not filled. These failed TSVs not only decrease the assembly yield

5.5 Issue of TSV-Based DRAM

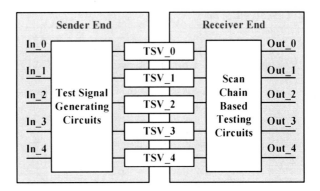

Fig. 5.13 TSV check

but also increases the total cost. This process problem can be solved by circuit design, such as TSV check and repair method.

First, this circuit is to check the TSV connectivity [7]. The objective of the testing circuits is to test all TSVs and locate the failed TSVs. On the sender end, circuits are added to generate test signals for each TSV and on the receiver end, scan chain based circuits are added to capture the test signals. It is shown in Fig. 5.13. Receiver end signal is compared with the sender end signal by using test signal generating circuits and scan chain based testing circuits. This internal circuit checks the success or failure of the TSV connectivity and help TSV repair.

This is the repair method [2]. In the Fig. 5.14, red X represents the failed TSVs. There are usually redundant TSVs for the TSV repair. In the conventional method, the redundant TSVs, r1 and r2, are dedicated and fixed. For this reason, the routing is very complex. The routing is marked in blue and the connections are decided by the MUX/DEMUX. However, in the proposed method, a failed TSV is repaired with a neighboring TSV. If a failure occurs at a TSV, the remaining TSVs are all shifted to the neighboring ones, whereby it is guaranteed that a failed TSV is always repaired with a neighboring TSV. This decreases the detour path, reduces routing complexity and loadings.

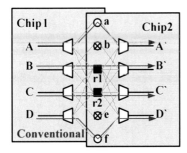 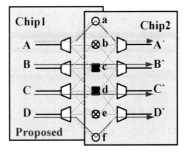

Fig. 5.14 TSV repair

References

1. J. Song et al., "An adaptive-bandwidth PLL for avoiding noise interference and DFE-less fast precharge sampling for over 10 Gb/s/pin graphics DRAM interface," in *IEEE Int. Solid-State Circuits Conf. Dig. Tech. Papers, 2013, pp. 312–313*.
2. U. Kang et al., "8 Gb 3D DDR3 DRAM using through-silicon-via technology," IEEE J. Solid-State Circuits, vol. 45, no. 1, pp. 111–119, Jan. 2010.
3. S.-B. Lim, H.-W. Lee, J. Song, and C. Kim, "A 247 μW 800 Mb/s/pin DLL-Based Data Self-Aligner for Through Silicon via (TSV) Interface," IEEE J. Solid-State Circuits, vol. 48, no. 3, pp. 711–723, Mar. 2013.
4. T. Kuroda et al., "A 0.9-V, 150-MHz, 10-mW, 4 mm2, 2-D discrete cosine transform core processor with variable threshold-voltage (VT) scheme," IEEE J. Solid-State Circuits, vol. 31, no. 11, pp. 828–837, Nov. 1996.
5. J.-S. Kim et al., "A 1.2 V 12.8 GB/s 2 Gb mobile Wide-I/O DRAM with 4 × 128 I/Os using TSV-based stacking," IEEE J. Solid-State Circuits, vol. 47, no. 1, pp. 107–116, Jan. 2012.
6. D. Malta et al., "Integrated process for defect-free copper plating and chemical-mechanical polishing of through-silicon vias for 3D interconnects," in *ECTC*, pp. 1769–1775, 2010.
7. A-.C. Hsieh et al., "TSV redundancy: architecture and design issues in 3-D IC," *IEEE Trans. VLSI Systems*, pp. 711–722, Apr. 2012.

Index

A
ACT, 5
ACTIVE, 7
Adaptive bandwidth DLL, 37
Address bit inversion (ABI), 19

B
Bank, 5
Bank group, 16
Bird's beak, 31
Bit line sense amplifier (BLSA), 8
Boundary switching problem, 35
Burst chop function, 15
Burst length, 1

C
CAS Latency controller, 38
Channel compensation, 52
Chip Selector (CS), 5
Clock distribution, 45
Clock Enable (CKE), 5
Column Address Strobe (CAS), 5
    Latency, 38
Column to column delay, 4
Crosstalk, 60
Crosstalk equalizer, 62
Crosstalk-Induced-Jitter (CIJ), 51

D
Data Bus Inversion (DBI), 73, 74
Data confliction, 82
DBI, 20
DDR1, 13
DDR2, 13
DDR3, 14
DDR4, 16
De-emphasis, 53
Decision feedback equalization (DFE), 57
Die stacking package, 77

DIMM, 22
DLL, 25
DLL parameters, 43
DLL-based data self-aligner, 82
Double data rate (DDR), 1
DQS, 5
Duty cycle corrector, 41
Dynamic random access memory (DRAM), 1
Dynamic voltage/frequency scaling (DVFS), 8

E
Emphasis, 53
Equalization, 52

F
Far-end crosstalk (FEXT), 61
FFE, 57
Fly-by topology, 65

G
GDDR3, 20
GDDR5, 19
GIO_MUX, 17
Global I/O lines (GIO), 7
Global ZQ calibration, 72
Graphics DDR (GDDR), 1

H
Heterogeneous type, 80
Homogeneous type, 80
HSUL, 21

I
IDD, 9
IDD7, 9
Impedance calibration, 71
Impedance matching, 71
Impedance mismatching, 69
Input buffer, 66

Inter-symbol-interference (ISI), 51

L
LDQ, 5
Lock condition of DLL, 32
Lock range of DLL, 31
Lossy channel, 51
Low power DDR memory (LPDDR), 1
LPDDR2, 20
LPDDR3, 20

M
Merged dual coarse delay line (MDCDL), 37
Mode register set (MRS), 19
Multi chip package, 78

N
Near-end crosstalk (NEXT), 61
Negative delayed clock, 26

O
OCD, 18
ODT, 18
On Die Termination (ODT), 5

P
Phase mixer, 34
Power-down mode, 10
Pre-charge state, 4
Pre-emphasis, 53
Pre-fetch, 4
PRECHARGE, 7
Pseudo Open Drain (POD), 20

R
READ, 7
Refresh operation, 4
Register controlled DLL, 34
Row Address Strobe (RAS), 5 command, 38
RTO, 3
RTT, 19

S
Self-dynamic voltage scaling, 37
Sense amplifier, 2

Sensing Zero (SZ), 3
Simultaneous switching output (SSO), 19
Single data rate (SDR), 1
Skew, 62
Slew-rate calibration, 72
SSTL, 20
Staggered memory, 61
Strobe, 4
Synchronous dynamic random access memory (SDRAM), 1
Synchronous Mirror Delay (SMD), 32, 33

T
T-branch, 65
tCCD_L, 16
tCCD_S, 16
tCHGSH, 4
tCK, 28
TCSR, 18
tDC, 82
tDDLY, 28
tDECROW, 4
tDREP, 26
Telegrapher's equation, 69
Through Silicon via (TSV), 77
Timing diagram of DLL, 28
Training sequence, 66
Transmitter, 52
tRCD, 7
tRRD, 5
tSADEV, 4

U
UDQ, 5

V
Voltage of bit line pre-charge (VBLP), 3

W
WCK, 19
WRITE, 7
Write Enable (WE), 5

Z
ZQ, 5

Printed by Publishers' Graphics LLC
LMO131121.15.22.47